Lecture Notes Editorial Policies

Lecture Notes in Statistics provides a format for the informal and quick publication of monographs, case studies, and workshops of theoretical or applied importance. Thus, in some instances, proofs may be merely outlined and results presented which will later be published in a different form.

Publication of the Lecture Notes is intended as a service to the international statistical community, in that a commercial publisher, Springer-Verlag, can provide efficient distribution of documents that would otherwise have a restricted readership. Once published and copyrighted, they can be documented and discussed in the scientific literature.

Lecture Notes are reprinted photographically from the copy delivered in camera-ready form by the author or editor. Springer-Verlag provides technical instructions for the preparation of manuscripts. Volumes should be no less than 100 pages and preferably no more than 400 pages. A subject index is expected for authored but not edited volumes. Proposals for volumes should be sent to one of the series editors or addressed to "Statistics Editor" at Springer-Verlag in New York.

Authors of monographs receive 50 free copies of their book. Editors receive 50 free copies and are responsible for distributing them to contributors. Authors, editors, and contributors may purchase additional copies at the publisher's discount. No reprints of individual contributions will be supplied and no royalties are paid on Lecture Notes volumes. Springer-Verlag secures the copyright for each volume.

Series Editors:

Professor P. Bickel
Department of Statistics
University of California
Berkeley, California 94720
USA

Professor P. Diggle
Department of Mathematics
Lancaster University
Lancaster LA1 4YL
England

Professor S. Fienberg
Department of Statistics
Carnegie Mellon University
Pittsburgh, Pennsylvania 15213
USA

Professor K. Krickeberg
3 Rue de L'Estrapade
75005 Paris
France

Professor I. Olkin
Department of Statistics
Stanford University
Stanford, California 94305
USA

Professor N. Wermuth
Department of Psychology
Johannes Gutenberg University
Postfach 3980
D-6500 Mainz
Germany

Professor S. Zeger
Department of Biostatistics
The Johns Hopkins University
615 N. Wolfe Street
Baltimore, Maryland 21205-2103
USA

Lecture Notes in Statistics 176

Springer
New York
Berlin
Heidelberg
Hong Kong
London
Milan
Paris
Tokyo

Zehua Chen
Zhidong Bai
Bimal K. Sinha

Ranked Set Sampling

Theory and Applications

Springer

Zehua Chen
Department of Statistics
and Applied Probability
National University of Singapore
3 Science Drive 2
Singapore 117543
Republic of Singapore
stachenz@nus.edu.sg

Zhidong Bai
Department of Statistics
and Applied Probability
National University of Singapore
3 Science Drive 2
Singapore 117543
Republic of Singapore
stabaizd@leonis.nus.edu.sg

Bimal K. Sinha
Department of Mathematics and Statistics
University of Maryland Baltimore County
1000 Hilltop Circle
Baltimore, MD 21250
USA
sinha@umbc.edu

Library of Congress Cataloging-in-Publication Data
Chen, Zehua.
Ranked set sampling : theory and applications / Zehua Chen, Zhidong Bai, Bimal K. Sinha.
p. cm. — (Lecture notes in statistics ; v. 176)
Includes bibliographical references and index.
ISBN 0-387-40263-2 (pbk. : alk. paper)
1. Sampling (Statistics) 2. Ranking and selection (Statistics) I. Bai, Zhidong. II. Sinha Bimal K., 1946- III. Title. IV. Lecture notes in statistics (Springer-Verlag) ; v. 176.
QA276.6.C447 2003
519.5'2—dc22 2003060810

ISBN 0-387-40263-2 Printed on acid-free paper.

Printed in the United States of America.

9 8 7 6 5 4 3 2 1 SPIN 10935187

Typesetting: Pages created by the authors using a Springer TEX macro package.

www.springer-ny.com

Springer-Verlag New York Berlin Heidelberg
A member of BertelsmannSpringer Science+Business Media GmbH

To my wife, Liang Shen, and my children,
Ruoxi, Xingjian, and Jingxuan

— Zehua Chen

To my wife, Dan Xicun, and sons, Li and Steve

— Zhidong Bai

To Suchandra Sinha, my loving wife and a caring
mother of our sons Biswajit and Satyajit

— Bimal K. Sinha

Preface

This monograph is the first book-length exposition of ranked set sampling. But, the subject matter is by no means new. The original notion of ranked set sampling, though not the technical term, was proposed by McIntyre in 1952. It was buried in the literature for quite a while. Its value was only re-discovered in recent years because of its cost-effective nature. Now, ranked set sampling has attracted practical interest in application areas such as agriculture, forestry, ecological and environmental science, and medical studies etc.. The theoretical foundations of the method has also been developed considerably, particularly during the last 15 years or so. A systematic exposition of the subject becomes necessary.

This book covers every development of RSS since the birth of the original idea. Statistical inferences based on RSS are investigated from the originally intended estimation of a population mean to many more complicated procedures such as the inferences on smooth-function-of-means, quantiles and density functions, the distribution-free tests and regression analyses. Various variants of the original RSS scheme are explored, including RSS with imperfect judgment ranking or ranking by concomitant variables, adaptive RSS, multi-layer RSS and a variety of unbalanced RSS with certain optimalities. Both parametric and non-parametric aspects of RSS are addressed. The book presents an equal mixture of theory and application. On one hand, theoretical foundations are provided with mathematical rigor, which a more serious reader can find beneficial. On the other hand, various methods are described in detail and illustrated by real or simulated data, which practitioners can find useful. In each chapter, the results of the chapter are summarized at the beginning so that the reader can see the whole picture at a glance. Sections which are mathematically involved or are only of technical interest are indicated with an asterisk. Though we intended to cover the topics in ranked set sampling systematically, the book is not meant to be encyclopedic. A considerable part of this book is based on our own research work. The selection of materials reflects our personal opinions. In order for the reader to have a more complete picture of the subject, bibliographical notes are given at the end of most of the chapters, which provide the sources of the materials presented in the book and relate to other works that are not presented or not discussed in detail. A comprehensive bibliography, which includes references both cited and not cited, is given at the end of the book.

The book can be used as a useful reference by the applied or theoretical statisticians as well as practitioners in various fields such as agriculture, sociology, ecological and environmental sciences, medical studies etc.. It can also be used as a text book for a short course at the graduate level.

Chapters 3 and 5 are written by Z. Bai. Chapter 8 and Sections 2.4, 2.9 are written by B. K. Sinha. The rest of the book is written by Z. Chen.

We would like to express our appreciation to M.S. Ridout for his detailed invaluable comments and suggestions on every chapter that resulted a substantial improvement of the first draft of this book. We would like to thank John Kimmel, the series editor, for his responsive handling of this book. Finally our gratitude goes to our families whose support was indispensable for this book to come into being.

Singapore — Zehua Chen
Baltimore, Maryland, USA — Zhidong Bai
June 2003 — Bimal K. Sinha

Contents

1

Introduction

Cost-effective sampling methods are of major concern in statistics, especially when the measurement of the characteristic of interest is costly and/or time-consuming. In the early 1950's, in seeking to effectively estimate the yield of pasture in Australia, McIntyre [96] proposed a sampling method which later became known as ranked set sampling (RSS). The notion of RSS provides an effective way to achieve observational economy under certain particular conditions. Although the method remained dormant for a long time, its value was rediscovered in the last 20 years or so because of its cost-effective nature. There is a surge of research interest of statisticians on RSS in recent years. There have been many new developments from the original idea of McIntyre, which made the method applicable in a much wider range of fields than originally intended. The theoretical foundation of RSS has been laid. Many variations and ramifications of the original idea have been proposed and studied. More and more applications of RSS have been cited in the literature. It is the purpose of this monograph to give a systematic account of the theory and application of RSS. In this introduction, we give a brief discussion on the notion of RSS and its applicability, a historical note on the development of RSS since it was proposed by McIntyre and, finally, an outline of the contents of the monograph.

1.1 What is ranked set sampling?

The basic premise for RSS is an infinite population under study and the assumption that a set of sampling units drawn from the population can be ranked by certain means rather cheaply *without* the actual measurement of the variable of interest which is costly and/or time-consuming. This assumption may look rather restrictive at first sight, but it turns out that there are plenty of situations in practice where this is satisfied. We shall give some examples at the end of this section.

Cycle 1

$$\begin{array}{l} \boldsymbol{X}_{(1)11} \le X_{(2)11} \le X_{(3)11} \Rightarrow \boldsymbol{X}_{(1)1} \\ X_{(1)21} \le \boldsymbol{X}_{(2)21} \le X_{(3)21} \Rightarrow \boldsymbol{X}_{(2)1} \\ X_{(1)31} \le X_{(2)31} \le \boldsymbol{X}_{(3)31} \Rightarrow \boldsymbol{X}_{(3)1} \end{array}$$

Cycle 2

$$\begin{array}{l} \boldsymbol{X}_{(1)12} \le X_{(2)12} \le X_{(3)12} \Rightarrow \boldsymbol{X}_{(1)2} \\ X_{(1)22} \le \boldsymbol{X}_{(2)22} \le X_{(3)22} \Rightarrow \boldsymbol{X}_{(2)2} \\ X_{(1)32} \le X_{(2)32} \le \boldsymbol{X}_{(3)32} \Rightarrow \boldsymbol{X}_{(3)2} \end{array}$$

... ...

Cycle m

$$\begin{array}{l} \boldsymbol{X}_{(1)1m} \le X_{(2)1m} \le X_{(3)1m} \Rightarrow \boldsymbol{X}_{(1)m} \\ X_{(1)2m} \le \boldsymbol{X}_{(2)2m} \le X_{(3)2m} \Rightarrow \boldsymbol{X}_{(2)m} \\ X_{(1)3m} \le X_{(2)3m} \le \boldsymbol{X}_{(3)3m} \Rightarrow \boldsymbol{X}_{(3)m} \end{array}$$

Fig. 1.1. Demonstration of the ranked set sampling procedure

The original form of RSS conceived by McIntyre can be described as follows. First, a simple random sample of size k is drawn from the population and the k sampling units are ranked with respect to the variable of interest, say X, by judgment *without* actual measurement. Then the unit with rank 1 is identified and taken for the measurement of X. The remaining units of the sample are discarded. Next, another simple random sample of size k is drawn and the units of the sample are ranked by judgment, the unit with rank 2 is taken for the measurement of X and the remaining units are discarded. This process is continued until a simple random sample of size k is taken and ranked and the unit with rank k is taken for the measurement of X. This whole process is referred to as a cycle. The cycle then repeats m times and yields a ranked set sample of size $N = mk$. For $k = 3$, the sampling procedure is illustrated in Figure 1.1.

The essence of RSS is conceptually similar to the classical stratified sampling. RSS can be considered as post-stratifying the sampling units according to their ranks in a sample. Although the mechanism is different from the stratified sampling, the effect is the same in that the population is divided into sub-populations such that the units within each sub-population are as homogeneous as possible. In fact, we can consider any mechanism, not necessarily ranking the units according to their X values, which can post-stratify the sampling units in such a way that it does not result in a random permutation of the units. The mechanism will then have similar effect to the ranking mechanism considered above. In the next chapter, we will discuss several such mechanisms. Therefore, we can extend the notion of RSS to a more general version.

We now give a general description of an RSS procedure. Suppose a total of N sampling units are to be measured on the variable of interest. First, N sets of units, each of size k, are drawn at random from the population under study. Then the units of each set are *ranked* by a certain mechanism other than using the actual measurements of the variable of interest. Finally, one and only one unit in each set with a pre-specified *rank* is measured on the variable of interest. It ends up with a sample of N measurements on the variable of interest which is to be referred to as a ranked set sample. Here, as mentioned earlier, it is implicitly assumed that the measurement of the variable of interest is costly and/or time-consuming but the ranking of a set of sampling units can be easily done with a negligible cost. It should be noted that, in the procedure, Nk units are sampled but only N of them are actually measured on the variable of interest. Let n_r be the number of measurements on units with rank r, $r = 1, \ldots, k$, such that $\sum_{r=1}^{k} n_r = N$. Denote by $X_{[r]i}$ the measurement on the ith measured unit with rank r. Then the ranked set sample can be represented as

$$\begin{array}{cccc} X_{[1]1} & X_{[1]2} & \cdots & X_{[1]n_1} \\ X_{[2]1} & X_{[2]2} & \cdots & X_{[2]n_2} \\ \cdots & \cdots & \cdots & \cdots \\ X_{[k]1} & X_{[k]2} & \cdots & X_{[k]n_k} \end{array} \tag{1.1}$$

If $n_1 = n_2 = \cdots = n_k$, the RSS is said to be balanced, otherwise, it is said to be unbalanced. The ranks which the units in a set receive may not necessarily tally with the numerical orders of their latent X values. If the ranks do tally with the numerical orders, the ranking is said to be perfect, otherwise, it is said to be imperfect. We use different notations to distinguish between perfect and imperfect ranking. For whatever quantities indexed by the ranks, we put the ranks in parentheses when ranking is perfect, otherwise, we put the ranks in brackets. Thus, $X_{(r)}$ and $X_{[r]}$ will be generic notation for measurements with rank r when ranking is perfect and imperfect, respectively.

RSS is applicable whenever a ranking mechanism can be found such that the ranking of sampling units is carried out easily and sampling is much cheaper than the measurement of the variable of interest. In particular, it is applicable in the following situations: (i) the ranking of a set of sampling units can be done easily by judgment relating to their latent values of the variable of interest through visual inspection or with the help of certain auxiliary means; (ii) there are certain easily obtainable concomitant variables available. By concomitant variables we mean variables that are not of major concern but are correlated with the variable of interest. These situations are abundant in practice. We mention a few examples in the following.

Example 1. RSS has been used in the assessments of animal carrying capacity of pastures and ranges. The sampling units in such assessments are well defined quadrats. The measurement of a quadrat involves mowing herbage and/or clipping browse within the quadrat and then drying and weighing the forage, which is quite time-consuming and destructive. However, nearby quadrats can be ranked rather precisely by visual inspection of an experienced person. If the variation between closely spaced quadrats is the same as that between widely spaced quadrats, quadrats taken nearby can be considered as random samples. RSS can increase greatly the efficiency in such situation. Halls and Dell [58] reported an experiment by using ranked set sampling for estimating forage yield in shortleaf-loblolly pine-hardwood forests carried out on a 300-acre tract of the Stephen F. Austin Experimental Forest, near Nacogdoches, Texas, USA. In this investigation, the quadrats were determined by metal frames of 3.1 feet square. The measurement of a quadrat involves clipping the browse and herbage on the quadrat and then dry the forage to constant weight at 70°C. Browse included the current season's leaf and twig growth of shrubs, woody vines, and trees available to deer or cattle. Herbage included all nonwoody plants. Due to particular features of the plantation in this area, the variation of the forage plants between closely spaced quadrats was essentially the same as that between widely spaced quadrats. The sampling procedure is as follows. First, set locations were established on a 2-chain grid. Then at each location, three metal frames were placed on the ground at randomly selected points within a circle 13 feet in radius. Two observers—one a professional range man, the other a woodsworker—independently ranked the three quadrats. Browse and herbage were ranked separately. To investigate the efficiency of the ranked set sampling, all the forage on all the quadrats were clipped and dried. For browse, 126 sets of three quadrats were ranked and clipped and, for herbage, 124 sets were ranked and clipped. It turned out that, to achieve the same accuracy of the estimation with 100 randomly selected qradrats, only 48 quadrats need to be measured for browse and only 73 quadrats need to be measured for herbage by RSS. For more applications in this area, see McIntyre [96] and Cobby et al. [48].

Example 2. Martin et al. [95] evaluated RSS for estimating shrub phytomass in Appalachian Oak forest. In this application, sampling involves counting the number of each vegetation type in randomly selected blocks of forest stands, which is rather time-consuming. But the ranking of a small number of blocks by visual inspection can be done quite easily.

In the above two examples, ranking of a small number of sampling units is done by visual inspection. For the visual inspection to be possible, the sampling units must be nearby. As mentioned in *Example 1*, if the variation between nearby units is the same as that between far apart units, a set of sampling units taken nearby can be considered as a random sample. However, if the variation between nearby units is much smaller than that between far apart units, a set taken nearby is no longer a random sample from the population under study. To overcome this difficulty, either the entire area under study may be divided into sections such that within each section local variations are essentially the same as the overall variation and then apply RSS in each section, or some auxiliary tools such as photos and video pictures may be used for the ranking of a randomly selected set of units.

Example 3. The assessment of the status of hazard waste sites is usually costly. But, often, a great deal of knowledge about hazard waste sites can be obtained from records, photos and certain physical characteristics, etc., and then be used to rank the hazard waste sites. In certain cases, the contamination levels of hazardous waste sites can be indicated either by visual cues such as defoliation or soil discoloration, or by inexpensive indicators such as special chemically-responsive papers, or electromagnetic readings. A concrete example follows. In the early 1970s, an investigation was carried out to estimate the total amount of plutonium $^{239,240}Pu$ in surface soil within a fenced area adjacent to the Nevada Test Site of Nevada, U.S.A. Soil samples at different locations were collected and analyzed for $^{239,240}Pu$ using radiochemical techniques to determine the plutonium concentration per square meter of surface soil. The radiochemical analysis of the soil samples is costly. However, a concomitant variable, the Field Instrument for the Determination of Low Energy Radiation (FIDLER) counts per minute taken at soil sample location, can be obtained rather cheaply. The ranking of soil sample locations based on the FIDLER is rather cheap and simple. See Gilbert and Eberhardt [56]. Yu and Lam [173] applied retrospectively RSS with concomitant variables to the data collected from the above investigation and found that the statistical procedures based on RSS improves significantly those based on simple random samples.

Example 4. In population census in certain countries, values of some variables are kept on a "short form" for all individuals. These records are available for a survey designer before a survey is carried out. In a survey, the values of survey variables are collected on a "long form". The variables on the "short form", which are easily obtainable, can be considered as concomitant variables. Therefore the RSS with concomitant variables can well be applied in surveys of this kind. The following case, which concerns the 1988 Test Population Census of Limeira, São Paulo, Brazil, provides an example. The test census was carried out in two stages. In the first stage, a population of about 44,000 households was censused with a "short form" questionnaire. In the second stage, a systematic sample of about 10% of the population size was surveyed with a "long form" questionnaire. The "short form" contains variables such as sex, age and education of the head of household, ownership of house, car and color TV, the number of rooms and bathrooms, a proxy to the monthly income of the head of household, etc.. The "long form" contains, besides the variables in the "short form", the actual monthly income of the head of household and other variables. The data obtained from the survey consists of the sample records of the "long form" for 426 heads of households. Details of the data were described in Silva and Skinner [152]. Chen [40] used the data to illustrate the efficiency of an adaptive RSS procedure and found that, for the estimation of the mean monthly income of the head of household, RSS is more efficient than simple random sampling in all statistical procedures considered. The relative efficiency in those procedures ranges from 1.068 to 1.24.

Example 5. Many quantitative traits of human such as hypertension and obesity can be attributed to genetic factors. In genetic linkage analysis, sib-pair models are used for mapping quantitative trait loci in humans. To test whether or not a marker locus is associated with the quantitative trait under consideration, sib pairs are selected, the values of the quantitative trait of the pairs are measured, the genotypes at the locus of the pair are determined and the number of alleles that the pair have derived from a common parent [identical by descent (ibd)] at the locus is found. The data is then used to test whether the number of shared alleles ibd of the pair is correlated with the squared difference between the values of the quantitative trait of the pair. However, the power of the test by using a simple random sample is very low. To detect the association in existence, thousands of sib pairs are usually required. Risch and Zhang [136] found that the power of the test can be significantly increased by selecting sib pairs with extremely concordant or disconcordant trait values. In implementation, this requires to screen a large number of sib pairs before genotyping can be started, which will certainly be subject to practical limitations. To overcome the difficulty caused by practical limitations, a unbalanced RSS scheme can be employed for the selection of the sib pairs. Sib pairs can be screened in sets of an appropriate size. In each set, only the pair with either the smallest or the largest absolute difference in trait values is selected for genetyping. This procedure, while increasing the power of the test, is more practical.

Example 6. RSS can be used in certain medical studies. For instance, it can be used in the determination of normal ranges of certain medical measures, which usually involves expensive laboratory tests. Samawi [140] considered using RSS for the determination of normal ranges of bilirubin level in blood for new born babies. To establish such ranges, blood sample must be taken from the sampled babies and tested in a laboratory. But, on the other hand, the ranking of the bilirubin levels of a small number of babies can be done by observing whether their face, chest, lower parts of the body and the terminal parts of the whole body are yellowish, since, as the yellowish color goes from face to the terminal parts of the whole body, the level of bilirubin in blood goes higher. RSS also has potential applications in clinical trials. Usually, the cost for a patient to go through a clinical trial is relatively high. However, the patients to be involved in the trial can be selected using the technique of RSS based on their information such as age, weight, height, blood pressure and health history etc., which can be obtained with a relatively negligible cost.

RSS is expected to be more efficient than simple random sampling (SRS), when applicable, regardless how ranking is done. This is because, intuitively, a ranked set sample contains more information than a simple random sample of the same size, since a ranked set sample contains not only the information carried by the measurements on the variable of interest but also the information carried by the ranks. Throughout this monograph, we will establish this fact in many different contexts. In certain situations, further improvement can be achieved by appropriately designed unbalanced RSS schemes. We will elaborate this point in later chapters. With its great potential of application in many practical fields, the method of RSS is becoming an important statistical tool kit.

1.2 A historical note

The idea of ranked set sampling was first proposed by McIntyre [96] in his effort to find a more efficient method to estimate the yield of pastures. Measuring yield of pasture plots requires mowing and weighing the hay which is time-consuming. But an experienced person can rank by eye inspection fairly accurately the yields of a small number of plots without actual measurement. McIntyre adopted the following sampling scheme. Each time, a random sample of k pasture lots is taken and the lots are ranked by eye inspection with respect to the amount of yield. From the first sample, the lot with rank 1 is taken for cutting and weighing. From the second sample, the lot with rank 2 is taken, and so on. When each of the ranks from 1 to k has an associated lot being taken for cutting and weighing, the cycle repeats over again and again until a total of m cycles are completed. McIntyre illustrated the gain in efficiency by a computation involving five distributions. He observed that the relative efficiency, defined as the ratio of the variance of the mean of a simple random sample and the variance of the mean of a ranked set sample of the same size, is not much less than $(k+1)/2$ for symmetric or moderately asymmetric distributions, and that the relative efficiency diminishes with increasing asymmetry of the underlying distribution but is always greater than 1. McIntyre also illustrated the estimation of higher moments. In addition, McIntyre mentioned the problem of optimal allocation of the measurements among the ranks and the problems of ranking error and possible correlation among the units within a set, etc. Though there is no theoretical rigor, the work of McIntyre [96] is pioneering and fundamental since the spores of many later developments of RSS are already contained in this paper.

The idea of RSS seemed buried in the literature for a long time until Halls and Dell [58] conducted a field trial evaluating its applicability to the estimation of forage yields in a pine hardwood forest. The terminology *ranked set sampling* was, in fact, coined by Halls and Dell. The first theoretical result about RSS was obtained by Takahasi and Wakimoto [167]. They proved that, when ranking is perfect, the ranked set sample mean is an unbiased estimator of the population mean , and the variance of the ranked set sample mean is always smaller than the variance of the mean of a simple random sample of the same size. Dell and Clutter [50] later obtained similar results, however, without restricting to the case of perfect ranking. Dell and Clutter [50] and David and Levine [49] were the first to give some theoretical treatments on imperfect ranking. Stokes [156] [157] considered the use of concomitant variables in RSS. Up to this point, the attention had been focused mainly on the nonparametric estimation of population mean. A few years later, Stokes [159]

[158] considered the estimation of population variance and the estimation of correlation coefficient of a bivariate normal population based on an RSS. However, other statistical procedures and new methodologies in the context of RSS had yet to be investigated and developed.

The middle of 1980's was a turning point in the development of the theory and methodology of RSS. Since then, various statistical procedures with RSS, non-parametric or parametric, have been investigated, variations of the original notion of RSS have been proposed and developed, and sound general theoretical foundations of RSS have been laid. A few references of these developments are given below. The estimation of cumulative distribution function with various settings of RSS was considered by Stokes and Sager [162], Kvam and Samaniego [83] and Chen [35]. The RSS version of distribution-free test procedures such as sign test, signed rank test and Mann-Whitney-Wilconxon test were investigated by Bohn and Wolfe [27] [28], and Hettmansperger [60]. The estimation of density function and population quantiles using RSS data were studied by Chen [34] [35]. The RSS counterpart of ratio estimate was considered by Samawi and Muttlak [143]. The U-statistic and M-estimation based on RSS were considered, respectively, by Presnell and Bohn [134] and Zhao and Chen [175]. The RSS regression estimate was tackled by Patil et al. [125], Yu and Lam [173] and Chen [38]. The parametric RSS assuming the knowledge of the family of the underlying distribution was studied by many authors, e.g., Abu-Dayyeh and Muttlak [1], Bhoj [19], Bhoj and Ahsanullah [20], Fei et al. [55], Li and Chuiv [91], Shen [147], Sinha et al. [155], Stokes [161], Chen [36]. The optimal design in the context of unbalanced RSS was considered by Stokes [161], Kaur et al. [74] [70] [71], Ozturk and Wolfe [116], Chen and Bai [41] and Chen [39]. A general theory on parametric and non-parametric RSS was developed by Bai and Chen [6]. Ranking mechanisms based on the use of multiple concomitant variables were developed by Chen and Shen [42] and Chen [40]. For an extensive bibliography of RSS before 1999, the reader is referred to Patil et al. [130].

1.3 Scope of the rest of the monograph

Chapter 2 discusses balanced non-parametric RSS. We start with the original form of McIntyre and consider the estimation of the mean and variance of a population. These results are used for developing suitable tests for the population mean. Then we generalize the original form and develop a general theory on the non-parametric RSS. We also consider the estimation of cumulative distribution function, probability density function and population quantiles. The M-estimate is also dealt with in this chapter.

In Chapter 3, balanced parametric RSS is discussed. First, we lay a theoretical foundation for the parametric RSS by considering the Fisher information. Then the maximum likelihood estimation based on RSS is considered and the best linear unbiased estimate for location-scale family distributions is discussed. An appendix gives some technical details.

Chapter 4 is devoted to unbalanced RSS. First, the methodology of analyzing unbalanced ranked set data is developed for the inferences on distribution functions, quantiles and general statistical functionals including mean, variance, etc.. Then the chapter is focused on the methodology of the design of optimal unbalanced RSS. The optimal designs for parametric location-scale families and for non-parametric estimation of quantiles are treated in detail. The methods of Bayes design and adaptive design are briefly discussed.

Chapter 5 deals with classical distribution-free tests in the context of RSS. The sign test, signed rank test and Mann-Whitney-Wilcoxon test are considered. The design of optimal RSS for distribution-free tests is also discussed.

Chapter 6 tackles the ranked set sampling with concomitant variables. A multi-layer RSS scheme and an adaptive RSS scheme using multiple concomitant variables are developed. The regression-type estimate of a population mean and the general regression analysis using RSS are discussed. The design of optimal RSS schemes for regression analysis using concomitant variables is explored.

Chapter 7 explores the application of the RSS techniques to data reduction for data mining.

The monograph ends with illustrations of several case studies in Chapter 8.

2

Balanced Ranked Set Sampling I: Nonparametric

In a balanced ranked set sampling, the number of measurements made on each ranked statistic is the same for all the ranks. A balanced ranked set sampling produces a data set as follows.

$$\begin{array}{cccc} X_{[1]1} & X_{[1]2} & \cdots & X_{[1]m} \\ X_{[2]1} & X_{[2]2} & \cdots & X_{[2]m} \\ \cdots & \cdots & \cdots & \cdots, \\ X_{[k]1} & X_{[k]2} & \cdots & X_{[k]m}. \end{array} \tag{2.1}$$

It should be noted that the $X_{[r]i}$'s in (2.1) are all mutually independent and, in addition, the $X_{[r]i}$'s in the same row are identically distributed. We denote by $f_{[r]}$ and $F_{[r]}$, respectively, the density function and the distribution function of the common distribution of the $X_{[r]i}$'s. The density function and the distribution function of the underlying distribution are denoted, respectively, by f and F. In this chapter, no assumption on the underlying distribution is made. We shall first discuss the ranking mechanisms of RSS and then discuss the properties of various statistical procedures using ranked set samples. In Section 1, the concept of consistent ranking mechanism is proposed and several consistent ranking mechanisms are discussed. In Section 2.2, the estimation of means using RSS sample is considered. The unbiasedness, asymptotic distribution and relative efficiency of the RSS estimates with respect to the corresponding SRS estimates are treated in detail. In Section 2.3, the results on the estimation of means are extended to the estimation of smooth functions of means. In Section 2.4, a special treatment is devoted to the estimation of variance. A minimum variance unbiased non-negative estimate of variance based on an RSS sample is proposed and studied. In Section 2.5, tests and confidence interval procedures for the population mean are discussed. In Section 2.6, the estimation of quantiles is tackled. The RSS sample quantiles are defined and their properties, similar to those of the SRS sample quantiles, such as strong consistency, Bahadur representation and asymptotic normality are established. In Section 2.7, the estimation of density function is treated. A kernel estimate based on an RSS sample is defined similarly as in the case of SRS. Its properties and relative efficiency with respect to SRS are investigated. In Section 2.8, the properties of M-estimates using RSS samples are discussed. Some technical details on the estimation of variance are given in Section 2.9. Readers who are not interested in technicalities might skip the final section.

2.1 Ranking mechanisms

In chapter 1, we introduced the general procedure of RSS. The procedure is a two-stage scheme. At the first stage, simple random samples are drawn and a certain ranking mechanism is employed to rank the units in each simple random sample. At the second stage, actual measurements of the variable of interest are made on the units selected based on the ranking information obtained at the first stage. The judgment ranking relating to the latent values of the variable of interest, as originally considered by McIntyre [96], provides one ranking mechanism. We mentioned that mechanisms other than this one can be used as well. In this section, we discuss the ranking mechanisms which are to be used in various practical situations.

Let us start with McIntyre's original ranking mechanism, i.e., ranking with respect to the latent values of the variable of interest. If the ranking is perfect, that is, the ranks of the units tally with the numerical orders of their latent values of the variable of interest, the measured values of the variable of interest are indeed order statistics. In this case, $f_{[r]} = f_{(r)}$, the density function of the rth order statistic of a simple random sample of size k from distribution F. We have

$$f_{(r)}(x) = \frac{k!}{(r-1)!(k-r)!} F^{r-1}(x)[1 - F(x)]^{k-r} f(x).$$

It is then easy to verify that

$$f(x) = \frac{1}{k} \sum_{r=1}^{k} f_{(r)}(x),$$

for all x. This equality plays a very important role in RSS. It is this equality that gives rise to the merits of RSS. We are going to refer to equalities of this kind as fundamental equalities.

A ranking mechanism is said to be consistent if the following fundamental equality holds:

$$F(x) = \frac{1}{k} \sum_{r=1}^{k} F_{[r]}(x), \text{ for all } x. \tag{2.2}$$

Obviously, perfect ranking with respect to the latent values of X is consistent. We discuss other consistent ranking mechanisms in what follows.

(i) Imperfect ranking with respect to the variable of interest. When there are ranking errors, the density function of the ranked statistic with rank r is no longer $f_{(r)}$. However, we can express the corresponding cumulative distribution function $F_{[r]}$ in the form:

$$F_{[r]}(x) = \sum_{s=1}^{k} p_{sr} F_{(s)}(x),$$

where p_{sr} denotes the probability with which the sth (numerical) order statistic is judged as having rank r. If these error probabilities are the same within each cycle of a balanced RSS, we have $\sum_{s=1}^{k} p_{sr} = \sum_{r=1}^{k} p_{sr} = 1$. Hence,

$$\begin{aligned}\frac{1}{k}\sum_{r=1}^{k} F_{[r]}(x) &= \frac{1}{k}\sum_{r=1}^{k}\sum_{s=1}^{k} p_{sr}F_{(s)}(x)\\ &= \frac{1}{k}\sum_{s=1}^{k}(\sum_{r=1}^{k} p_{sr})F_{(s)}(x) = F(x).\end{aligned}$$

(ii) Ranking with respect to a concomitant variable. There are cases in practical problems where the variable of interest, X, is hard to measure and difficult to rank as well but a concomitant variable, Y, can be easily measured. Then the concomitant variable can be used for the ranking of the sampling units. The RSS scheme is adapted in this situation as follows. At the first stage of RSS, the concomitant variable is measured on each unit in the simple random samples, and the units are ranked according to the numerical order of their values of the concomitant variable. Then the measured X values at the second stage are induced order statistics by the order of the Y values. Let $Y_{(r)}$ denote the rth order statistic of the Y's and $X_{[r]}$ denote its corresponding X. Let $f_{X|Y_{(r)}}(x|y)$ denote the conditional density function of X given $Y_{(r)} = y$ and $g_{(r)}(y)$ the marginal density function of $Y_{(r)}$. Then we have

$$f_{[r]}(x) = \int f_{X|Y_{(r)}}(x|y)g_{(r)}(y)dy.$$

It is easy to see that

$$\begin{aligned}f(x) &= \int \sum_{r=1}^{k} \frac{1}{k} f_{X|Y_{(r)}}(x|y)g_{(r)}(y)dy\\ &= \frac{1}{k}\sum_{r=1}^{k} f_{[r]}(x).\end{aligned}$$

(iii) Multivariate samples obtained by ranking one of the variables. Without loss of generality, let us consider the bivariate case. Suppose that inferences are to be made on the joint distribution of X and Y. The RSS scheme can

be similarly adapted in this case. The scheme goes the same as the standard RSS. The sampling units are ranked according to one of the variables, say Y. However, for each item to be quantified, both variables are measured. Let $f(x,y)$ denote the joint density function of X and Y and $f_{[r]}(x,y)$ the joint density function of $X_{[r]}$ and $Y_{[r]}$. Then

$$f_{[r]}(x,y) = f_{X|Y_{[r]}}(x|y)g_{[r]}(y)$$

and

$$f(x,y) = \frac{1}{k}\sum_{r=1}^{k} f_{[r]}(x,y).$$

(iv) Ranking mechanisms based on multiple concomitant variables. If there is more than one concomitant variable, any function of the concomitant variables can be used as a ranking criterion and the resultant ranking mechanism is consistent. Some of the ranking mechanisms based on functions of concomitant variables are discussed in detail in Chapter 6. We also develop a multi-layer RSS scheme using multiple concomitant variables, which is consistent, in Chapter 6.

2.2 Estimation of means using ranked set sample

Let $h(x)$ be any function of x. Denote by μ_h the expectation of $h(X)$, i.e., $\mu_h = Eh(X)$. We consider in this section the estimation of μ_h by using a ranked set sample. Examples of $h(x)$ include: (a) $h(x) = x^l, l = 1, 2, \cdots$, corresponding to the estimation of population moments, (b) $h(x) = I\{x \le c\}$ where $I\{\cdot\}$ is the usual indicator function, corresponding to the estimation of distribution function, (c) $h(x) = \frac{1}{\lambda}K(\frac{t-x}{\lambda})$, where K is a given function and λ is a given constant, corresponding to the estimation of density function. We assume that the variance of $h(X)$ exists. Define the moment estimator of μ_h based on data (2.1) as follows.

$$\hat{\mu}_{h\cdot\text{RSS}} = \frac{1}{mk}\sum_{r=1}^{k}\sum_{i=1}^{m} h(X_{[r]i}).$$

We consider first the statistical properties of $\hat{\mu}_{h\cdot\text{RSS}}$ and then the relative efficiency of RSS with respect to SRS in the estimation of means.

First, we have the following result.

Theorem 2.1. *Suppose that the ranking mechanism in RSS is consistent. Then,*

(i) The estimator $\hat{\mu}_{h\cdot RSS}$ is unbiased, i.e., $E\hat{\mu}_{h\cdot RSS} = \mu_h$.
(ii) $Var(\hat{\mu}_{h\cdot RSS}) \le \frac{\sigma_h^2}{mk}$, where σ_h^2 denotes the variance of $h(X)$, and the inequality is strict unless the ranking mechanism is purely random.

(iii) As $m \to \infty$,

$$\sqrt{mk}(\hat{\mu}_{h\cdot\text{RSS}} - \mu_h) \to N(0, \sigma^2_{h\cdot\text{RSS}}),$$

in distribution, where,

$$\sigma^2_{h\cdot\text{RSS}} = \frac{1}{k}\sum_{r=1}^{k}\sigma^2_{h[r]}.$$

Here $\sigma^2_{h[r]}$ denotes the variance of $h(X_{[r]i})$.

Proof: (i) It follows from the fundamental equality that

$$\begin{aligned} E\hat{\mu}_{h\cdot\text{RSS}} &= \frac{1}{mk}\sum_{r=1}^{k}\sum_{i=1}^{m} Eh(X_{[r]i}) = \frac{1}{k}\sum_{r=1}^{k} Eh(X_{[r]1}) \\ &= \frac{1}{k}\sum_{r=1}^{k}\int h(x)dF_{[r]}(x) = \int h(x)d\frac{1}{k}\sum_{r=1}^{k}F_{[r]}(x) \\ &= \int h(x)dF(x) = \mu_h. \end{aligned}$$

(ii)

$$\begin{aligned} \text{Var}(\hat{\mu}_{h\cdot\text{RSS}}) &= \frac{1}{(mk)^2}\sum_{r=1}^{k}\sum_{i=1}^{m}\text{Var}(h(X_{[r]i})) = \frac{1}{mk^2}\sum_{r=1}^{k}\text{Var}(h(X_{[r]})) \\ &= \frac{1}{mk}\left(\frac{1}{k}\sum_{r=1}^{k}(E[h(X_{[r]})]^2 - [Eh(X_{[r]})]^2)\right) \\ &= \frac{1}{mk}\left(m_{h2} - \frac{1}{k}\sum_{r=1}^{k}[Eh(X_{[r]})]^2\right), \end{aligned}$$

where m_{h2} denotes the second moment of $h(X)$. It follows from the Caushy-Schwarz inequality that

$$\frac{1}{k}\sum_{r=1}^{k}[Eh(X_{[r]})]^2 \geq \left(\frac{1}{k}\sum_{r=1}^{k}Eh(X_{[r]})\right)^2 = \mu_h^2,$$

where the equality holds only when $Eh(X_{[1]}) = \cdots = Eh(X_{[r]})$ in which case the ranking mechanism is purely random.

(iii) By the fundamental equality, $\mu_h = \frac{1}{k}\sum_{r=1}^{k}\mu_{h[r]}$, where $\mu_{h[r]}$ is the expectation of $h(X_{[r]i})$. Then, we can write

$$\begin{aligned} \sqrt{mk}(\hat{\mu}_{h\cdot\text{RSS}} - \mu_h) &= \frac{1}{\sqrt{k}}\sum_{r=1}^{k}\sqrt{m}[\frac{1}{m}\sum_{i=1}^{m}h(X_{[r]i}) - \mu_{h[r]}] \\ &= \frac{1}{\sqrt{k}}\sum_{r=1}^{k}Z_{mr}, \text{ say.} \end{aligned}$$

By the multivariate central limit theorem, $(Z_{m1}, \cdots, Z_{mk})$ converges to a multivariate normal distribution with mean vector zero and covariance matrix given by $Diag(\sigma^2_{h[1]}, \cdots, \sigma^2_{h[k]})$. Part (iii) then follows.

This proves the theorem.

We know that $\sigma^2_h/(mk)$ is the variance of the moment estimator of μ_h based on a simple random sample of size mk. Theorem 2.1 implies that the moment estimator of μ_h based on an RSS sample always has a smaller variance than its counterpart based on an SRS sample of the same size. In the context of RSS, we have tacitly assumed that the cost or effort for drawing sampling units from the population and then ranking them is negligible. When we compare the efficiency of a statistical procedure based on an RSS sample with that based on an SRS sample, we assume that the two samples have the same size. Let $\hat{\mu}_{h\cdot\text{SRS}}$ denote the sample mean of a simple random sample of size mk. We define the relative efficiency of RSS with respect to SRS in the estimation of μ_h as follows:

$$\text{RE}(\hat{\mu}_{h\cdot\text{RSS}}, \hat{\mu}_{h\cdot\text{SRS}}) = \frac{\text{Var}(\hat{\mu}_{h\cdot\text{SRS}})}{\text{Var}(\hat{\mu}_{h\cdot\text{RSS}})}. \tag{2.3}$$

Theorem 2.1 implies that $\text{RE}(\hat{\mu}_{h\cdot\text{RSS}}, \hat{\mu}_{h\cdot\text{SRS}}) \geq 1$. In order to investigate the relative efficiency in more detail, we derive the following:

$$\begin{aligned}
\sigma^2_{h\cdot\text{RSS}} &= \frac{1}{k}\sum_{r=1}^{k}\sigma^2_{h[r]} \\
&= \frac{1}{k}\sum_{r=1}^{k}(E[h(X_{[r]})]^2 - [Eh(X_{[r]})]^2) \\
&= \frac{1}{k}\sum_{r=1}^{k}E[h(X_{[r]})]^2 - \mu_h^2 + \mu_h^2 - \frac{1}{k}\sum_{r=1}^{k}[Eh(X_{[r]})]^2 \\
&= \sigma^2_h - \frac{1}{k}\sum_{r=1}^{k}(\mu_{h[r]} - \mu_h)^2. \qquad (2.4)
\end{aligned}$$

Thus, we can express the relative efficiency as

$$\text{RE}(\hat{\mu}_{h\cdot\text{RSS}}, \hat{\mu}_{h\cdot\text{SRS}}) = \frac{\sigma^2_h}{\sigma^2_{h\cdot\text{RSS}}} = \left[1 - \frac{\frac{1}{k}\sum_{r=1}^{k}(\mu_{h[r]} - \mu_h)^2}{\sigma^2_h}\right]^{-1}.$$

It is clear from the above expression that, as long as there is at least one r such that $\mu_{h[r]} \neq \mu_h$, the relative efficiency is greater than 1. For a given underlying distribution and a given function h, the relative efficiency can be computed, at least, in principle.

In the remainder of this section, we discuss the relative efficiency in more detail for the special case that $h(x) = x$. Based on the computations on a number of underlying distributions, McIntyre [96] made the following conjecture: the relative efficiency of RSS with respect to SRS, in the estimation

Table 2.1. $\mu, \sigma^2, \gamma, \kappa$ and the relative efficiency of RSS with $k = 2, 3, 4$ for some distributions

Distribution	μ	σ^2	γ	κ	2	3	4
Uniform	0.500	0.083	0.000	1.80	1.500	2.000	2.500
Exponential	1.000	1.000	2.000	9.00	1.333	1.636	1.920
Gamma(0.5)	0.500	0.500	2.828	15.0	1.254	1.483	1.696
Gamma(1.5)	1.500	1.500	1.633	7.00	1.370	1.710	2.030
Gamma(2.0)	2.000	2.000	1.414	6.00	1.391	1.753	2.096
Gamma(3.0)	3.000	3.000	1.155	5.00	1.414	1.801	2.169
Gamma(4.0)	4.000	4.000	1.000	4.50	1.427	1.827	2.210
Gamma(5.0)	5.000	5.000	0.894	4.20	1.434	1.843	2.236
Normal	0.000	1.000	0.000	3.00	1.467	1.914	2.347
Beta (4,4)	0.500	0.028	0.000	2,45	1.484	1.958	2.425
Beta(7,4)	0.636	0.019	-0.302	2.70	1.475	1.936	2.389
Beta(13,4)	0.765	0.010	-0.557	3.14	1.460	1.903	2.333
Weibull(0.5)	2.000	20.00	6.619	87.7	1.127	1.236	1.334
Weibull(1.5)	0.903	0.376	1.072	4.39	1.422	1.822	2.205
Weibull(2.0)	0.886	0.215	0.631	3.24	1.458	1.897	2.325
Weibull(3.0)	0.893	0.105	0.168	2.73	1.476	1.936	2.387
Weibull(4.0)	0.906	0.065	-0.087	2.75	1.474	1.932	2.380
Weibull(5.0)	0.918	0.044	-0.254	2.88	1.469	1.921	2.361
Weibull(6.0)	0.918	0.032	-0.373	3.04	1.464	1.909	2.341
Weibull(7.0)	0.935	0.025	-0.463	3.19	1.459	1.898	2.324
Weibull(8.0)	0.942	0.019	-0.534	3.33	1.456	1.890	2.309
$\chi^2(1)$	0.789	0.363	0.995	3.87	1.430	1.841	2.239
Triangular	0.500	0.042	0.000	2.40	1.485	1.961	2.430
Extreme value	0.577	1.645	1.300	5.40	1.413	1.793	2.153

of population mean, is between 1 and $(k+1)/2$ where k is the set size; For symmetric underlying distributions, the relative efficiency is not much less than $(k+1)/2$, however, as the underlying distribution becomes asymmetric, the relative efficiency drops down. Takahasi and Wakimoto [167] showed that, when ranking is perfect, $\frac{1}{k}\sum_{r=1}^{k}\sigma^2_{h[r]}$, as a function of k, decreases as k increases, which implies that the relative efficiency increases as k increases. A practical implication of this result is that, in the case of judgment ranking relating to the latent values of the variable of interest, when ranking accuracy can still be assured or, in other cases, when the cost of drawing sampling units and ranking by the given mechanism can still be kept at a negligible level, the set size k should be taken as large as possible. Dell and Clutter [50] computed the relative efficiency for a number of underlying distributions. They noticed that the relative efficiency is affected by the underlying distribution, especially by the skewness and kurtosis. Table 2.1 below is partially reproduced from Table 1 of Dell and Clutter (1972). The notations μ, σ^2, γ and κ in the table stand, respectively, for the mean, variance, skewness and kurtosis.

2.3 Estimation of smooth-function-of-means using ranked set sample

In this section we deal with the properties of RSS for a particular model, the smooth-function-of-means model, which refers to the situation where we are interested in the inference on a smooth function of population moments. Typical examples of smooth-function-of-means are (i) the variance, (ii) the coefficient of variation, and (iii) the correlation coefficient, etc. Let $m_1, \ldots, m_p$ denote p moments of F and g a p-variate smooth function with first derivatives. We consider the method-of-moment estimation of $g(m_1, \ldots, m_p)$.

The following notation will be used in this section. Let $Z_l, l = 1, \ldots, p$, be functions of $X(\sim F)$ such that $E[Z_l] = m_l$. Let $n = km$. A simple random sample of size n is represented by $\{(Z_{1j}, ..., Z_{pj}) : j = 1, \ldots, n\}$. A general RSS sample of size n is represented by $\{(Z_{1(r)i}, \ldots, Z_{p(r)i}) : r = 1, \ldots, k; i = 1, \ldots, m\}$. The simple random and ranked set sample moments are denoted, respectively, by

$$\bar{Z}_l = \frac{1}{n}\sum_{j=1}^{n} Z_{lj}, l = 1, \ldots, p$$

and

$$\tilde{Z}_l = \frac{1}{km}\sum_{r=1}^{k}\sum_{i=1}^{m} Z_{l(r)i}, l = 1, \ldots, p.$$

Let $\bar{\boldsymbol{Z}}_{\text{SRS}} = (\bar{Z}_1, \ldots, \bar{Z}_p)^T$ and $\bar{\boldsymbol{Z}}_{\text{RSS}} = (\tilde{Z}_1, \ldots, \tilde{Z}_p)^T$. Denote by Σ_{SRS} and Σ_{RSS} the variance-covariance matrices of $\sqrt{n}\bar{\boldsymbol{Z}}_{\text{SRS}}$ and $\sqrt{n}\bar{\boldsymbol{Z}}_{\text{RSS}}$, respectively. Let $\boldsymbol{\partial g}$ denote the vector of the first partial derivatives of g evaluated at $(m_1, \ldots, m_p)$. Define

$$\begin{aligned}\eta &= g(m_1, \ldots, m_p),\\ \hat{\eta}_{\text{SRS}} &= g(\bar{Z}_1, \ldots, \bar{Z}_p),\\ \hat{\eta}_{\text{RSS}} &= g(\tilde{Z}_1, \ldots, \tilde{Z}_p).\end{aligned}$$

We first state the asymptotic normality of $\hat{\eta}_{\text{SRS}}$ and $\hat{\eta}_{\text{RSS}}$, and then consider the asymptotic relative efficiency (ARE) of $\hat{\eta}_{\text{RSS}}$ with respect to $\hat{\eta}_{\text{SRS}}$.

Theorem 2.2. *As $m \to \infty$ (hence $n \to \infty$), we have*

$$\sqrt{n}(\hat{\eta}_{SRS} - \eta) \to N(0, \boldsymbol{\partial g}^T \Sigma_{SRS} \boldsymbol{\partial g})$$

in distribution and

$$\sqrt{n}(\hat{\eta}_{RSS} - \eta) \to N(0, \boldsymbol{\partial g}^T \Sigma_{RSS} \boldsymbol{\partial g})$$

in distribution.

The above result follows from the multivariate central limit theorem. The proof is omitted. The ARE of $\hat{\eta}_{\text{RSS}}$ with respect to $\hat{\eta}_{\text{SRS}}$ is defined as

$$\text{ARE}(\hat{\eta}_{\text{RSS}}, \hat{\eta}_{\text{SRS}}) = \frac{\boldsymbol{\partial g}^T \Sigma_{\text{SRS}} \boldsymbol{\partial g}}{\boldsymbol{\partial g}^T \Sigma_{\text{RSS}} \boldsymbol{\partial g}}.$$

The next theorem implies that the ARE of $\hat{\eta}_{\text{RSS}}$ with respect to $\hat{\eta}_{\text{SRS}}$ is always greater than 1.

Theorem 2.3. *Suppose that the ranking mechanism in RSS is consistent. Then we have that*

$$\Sigma_{SRS} \geq \Sigma_{RSS},$$

where $\Sigma_{SRS} \geq \Sigma_{RSS}$ means that $\Sigma_{SRS} - \Sigma_{RSS}$ is non-negative definite.

Proof: It suffices to prove that, for any vector of constants $\boldsymbol{a}$,

$$\boldsymbol{a}^T \Sigma_{\text{SRS}} \boldsymbol{a} - \boldsymbol{a}^T \Sigma_{\text{RSS}} \boldsymbol{a} \geq 0. \tag{2.5}$$

Define

$$Y = \boldsymbol{a}^T \boldsymbol{Z} = \sum_{j=1}^{p} a_j Z_j \quad \text{and} \quad \mu_Y = \boldsymbol{a}^T \mathbf{m} = \sum_{j=1}^{p} a_j m_j.$$

Then we have

$$\hat{\mu}_{Y \cdot \text{SRS}} = \boldsymbol{a}^T \bar{Z}_{\text{SRS}} \quad \text{and} \quad \hat{\mu}_{Y \cdot \text{RSS}} = \boldsymbol{a}^T \bar{Z}_{\text{RSS}}.$$

It follows from Theorem 2.1 that

$$\text{Var}(\hat{\mu}_{Y \cdot \text{RSS}}) \leq \text{Var}(\hat{\mu}_{Y \cdot \text{SRS}}),$$

i.e.,

$$\boldsymbol{a}^T \Sigma_{\text{SRS}} \boldsymbol{a} \geq \boldsymbol{a}^T \Sigma_{\text{RSS}} \boldsymbol{a}.$$

The theorem is proved.

In fact, it can be proved that, as long as there are at least two ranks, say r and s, such that $F_{[r]} \neq F_{[s]}$, then $\Sigma_{\text{SRS}} > \Sigma_{\text{RSS}}$.

It should be noted that, unlike in the estimation of means, the estimator of a smooth-function-of-means is no longer necessarily unbiased. It is only asymptotically unbiased. In this case, the relative efficiency of RSS with respect to SRS should be defined as the ratio of the mean square errors of the two estimators. The ARE, which is the limit of the relative efficiency as the samples size goes to infinity, does not take into account the bias for finite sample sizes. In general, the ARE can not be achieved when sample size is small. We consider this issue in more detail for the special case of the estimation of population variance σ^2 in the next section.

2.4 Estimation of variance using an RSS sample

2.4.1 Naive moment estimates

The natural estimates of σ^2 using an SRS sample and an RSS sample are given, respectively, by

$$s^2_{\text{SRS}} = \frac{1}{mk-1}\sum_{r=1}^{k}\sum_{i=1}^{m}(X_{ri} - \bar{X}_{\text{SRS}})^2,$$

where $\bar{X}_{\text{SRS}} = \frac{1}{mk}\sum_{r=1}^{k}\sum_{i=1}^{m} X_{ri}$, and

$$s^2_{\text{RSS}} = \frac{1}{mk-1}\sum_{r=1}^{k}\sum_{i=1}^{m}(X_{[r]i} - \bar{X}_{\text{RSS}})^2,$$

where $\bar{X}_{\text{RSS}} = \frac{1}{mk}\sum_{r=1}^{k}\sum_{i=1}^{m} X_{[r]i}$.

The properties of s^2_{RSS} were studied by Stokes [159]. Unlike the SRS version s^2_{SRS}, the RSS version s^2_{RSS} is biased. It can be derived, see Stokes [159], that

$$Es^2_{\text{RSS}} = \sigma^2 + \frac{1}{k(mk-1)}\sum_{r=1}^{k}(\mu_{[r]} - \mu)^2.$$

An appropriate measure of the relative efficiency of s^2_{RSS} with respect to s^2_{SRS} is then given by

$$\begin{aligned}\text{RE}(s^2_{\text{RSS}}, s^2_{\text{SRS}}) &= \frac{\text{Var}(s^2_{\text{SRS}})}{\text{MSE}(s^2_{\text{RSS}})} \\ &= \frac{\text{Var}(s^2_{\text{SRS}})}{\text{Var}(s^2_{\text{RSS}}) + \left[\frac{1}{k(mk-1)}\sum_{r=1}^{k}(\mu_{[r]} - \mu)^2\right]^2}.\end{aligned}$$

It can be easily seen that

$$\text{RE}(s^2_{\text{RSS}}, s^2_{\text{SRS}}) < \text{ARE}(s^2_{\text{RSS}}, s^2_{\text{SRS}}).$$

Since

$$\frac{1}{k}\sum_{r=1}^{k}(\mu_{[r]} - \mu)^2 < \sigma^2,$$

it is clear that $\frac{1}{k(mk-1)}\sum_{r=1}^{k}(\mu_{[r]} - \mu)^2$ will decrease as either k or m increases. That is, the RE will converge increasingly to the ARE as either k or m increases. Stokes [159] computed both the RE when $m = 1$ and the ARE for a few underlying distributions. Table 2.2 below is reproduced from Table 1 of [159]. In the table, the U-shaped distribution refers to the distribution with

Table 2.2. Relative efficiency, $\text{Var}(s^2_{\text{SRS}})/\text{MSE}(s^2_{\text{RSS}})$, for $m = 1$ and $m \to \infty$.

Distribution	k	$m = 1$	$m \to \infty$
(i) U-shaped	2	0.85	1.00
	3	1.22	1.10
	4	1.28	1.21
(ii) Uniform	2	0.72	1.00
	3	0.92	1.11
	4	1.09	1.25
	5	1.20	1.40
(iii) Normal	2	0.68	1.00
	3	0.81	1.08
	4	0.93	1.18
	5	1.03	1.27
(iv) Gamma	2	0.71	1.02
	3	0.81	1.08
	4	0.91	1.16
	5	1.00	1.23
	6	1.09	1.35
(v) Exponential	2	0.78	1.03
	3	0.84	1.08
	4	0.91	1.12
	5	0.97	1.17
	6	1.02	1.22
(vi) Lognormal	2	0.93	1.00
	3	0.95	1.01
	4	0.96	1.01
	5	0.97	1.02
	6	0.98	1.02
	7	0.99	1.03
	8	1.00	1.03
	9	1.01	1.04

density function $f(x) = (3/2)x^2 I\{-1 \leq x \leq 1\}$, and the Gamma distribution has density function $f(x) = x^4 \exp(-x)/\Gamma(5) I\{x \geq 0\}$.

It should be remarked that, in the estimation of variance, RSS is not necessarily more efficient than SRS when sample size is small, and the relative efficiency is much smaller than in the estimation of population mean even when RSS is beneficial. Therefore, if the estimation of variance is the primary purpose, it is not worthwhile to apply RSS. RSS is most useful when both the population mean and variance are to be estimated.

It is indeed a natural question whether or not better estimates of σ^2 based on an RSS sample can be found. We take up this question in the next subsection.

2.4.2 Minimum variance unbiased non-negative estimate

We demonstrate in this subsection that it is possible to construct a class of nonnegative unbiased estimates of σ^2 based on a balanced RSS sample, whatever the nature of the underlying distribution. Towards this end, we need the following basic identity which follows directly from (2.4):

$$\sigma^2 = \frac{1}{k}[\sum_{r=1}^{k} \sigma^2_{[r]} + \sum_{r=1}^{k} \mu^2_{[r]}] - \mu^2. \tag{2.6}$$

Recall that $\bar{X}_{i:\text{RSS}} = (1/k)\sum_{r=1}^{k} X_{[r]i}$ provides an unbiased estimate of the mean μ based on the data of the ith cycle of an RSS . Let

$$W_i = \sum_{r=1}^{k}(X_{[r]i} - \bar{X}_{i:\text{RSS}})^2, \quad i = 1, \cdots, m. \tag{2.7}$$

From the basic identity (2.6), it is clear that an unbiased estimate of σ^2 can be obtained by plugging in unbiased estimates of $\sigma^2_{[r]} + \mu^2_{[r]}$ and μ^2. Since $\sum_{i=1}^{m} X^2_{[r]i}/m$ is an unbiased estimate of the former term and $\bar{X}_{i:\text{RSS}}\bar{X}_{j:\text{RSS}}$ for $i \neq j$ is an unbiased estimate of μ^2, it follows easily that an unbiased estimate of σ^2 is given by

$$\hat{\sigma}^2 = \frac{\sum_{r=1}^{k}\sum_{i=1}^{m}[X_{[r]i}]^2}{mk} - \frac{\sum_{i \neq j} \bar{X}_{i:\text{RSS}}\bar{X}_{j:\text{RSS}}}{m(m-1)}. \tag{2.8}$$

The above estimate can be readily simplified as

$$\hat{\sigma}^2 = \frac{W}{mk} + \frac{B}{(m-1)k} \tag{2.9}$$

where B and W represent, respectively, the between- and within-cycle sum of squares of the entire balanced data, defined as

$$B = k\sum_{i=1}^{m}(\bar{X}_{i:\text{RSS}} - \bar{X}_{\text{RSS}})^2, \quad W = \sum_{i=1}^{m} W_i. \tag{2.10}$$

It is obvious that $\hat{\sigma}^2$ is nonnegative.

Let

$$\boldsymbol{X}_{[r]} = (X_{[r]1}, \cdots, X_{[r]m})', \; r = 1, \cdots, k,$$
$$\boldsymbol{X} = (\boldsymbol{X}'_{[1]}, \cdots, \boldsymbol{X}'_{[k]})'.$$

Denote by $\mathbf{1}_m$ an m-dimensional vector of elements 1's and I_m an identity matrix of order m. It can be easily verified that

$$\hat{\sigma}^2 = \sum_{r=1}^{k}\sum_{s=1}^{k}[\tilde{a}_{rs}\boldsymbol{X}_{[r]}^{'}(I_m - \frac{\mathbf{1}_m\mathbf{1}_m^{'}}{m})\boldsymbol{X}_{[s]} + \tilde{d}_{rs}\boldsymbol{X}_{[r]}^{'}\frac{\mathbf{1}_m\mathbf{1}_m^{'}}{m}\boldsymbol{X}_{[s]}],$$

where

$$\tilde{a}_{rs} = \begin{cases} \frac{1}{m(m-1)k^2}, & r \neq s, \\ \frac{(m-1)k+1}{m(m-1)k^2}, & r = s, \end{cases}$$

$$\tilde{d}_{rs} = \begin{cases} 1-m, & r \neq s, \\ 1-m+\frac{1}{mk}, & r = s. \end{cases}$$

Furthermore, let

$$\tilde{A} = (\tilde{a}_{rs})_{k\times k}, \quad \tilde{D} = (\tilde{d}_{rs})_{k\times k}.$$

We can express $\hat{\sigma}^2$ as follows:

$$\hat{\sigma}^2 = \boldsymbol{X}^{'}[\tilde{A}\otimes(I_m - \frac{\mathbf{1}_m\mathbf{1}_m^{'}}{m}) + \tilde{D}\otimes\frac{\mathbf{1}_m\mathbf{1}_m^{'}}{m}]\boldsymbol{X},$$

where $\otimes$ denotes the operation of Kronecker product.

The form of $\hat{\sigma}^2$ above motivates us to consider a class of quadratic estimators of σ^2. Let

$$Q = A\otimes(I_m - \frac{\mathbf{1}_m\mathbf{1}_m^{'}}{m}) + D\otimes\frac{\mathbf{1}_m\mathbf{1}_m^{'}}{m},$$

where $A = (a_{rs})_{k\times k}$ and $D = (d_{rs})_{k\times k}$ are arbitrary $k \times k$ symmetric matrices. We consider the class of quadratic estimators $\boldsymbol{X}^{'}Q\boldsymbol{X}$ with Q given by the above form. The following theorem provides characterizations of unbiased estimators of σ^2 and of the minimum variance unbiased estimator of σ^2 among estimators in the above class.

Theorem 2.4. *(a) $\hat{\sigma}^2 = \boldsymbol{X}^{'}Q\boldsymbol{X}$ in the above class is unbiased for σ^2 if and only if*

$$a_{rr} = \frac{1}{k(m-1)}[1 - \frac{k-1}{mk}] = \frac{1}{mk}[1 + \frac{1}{k(m-1)}]$$

for $r = 1, 2, \ldots, k$, and $D = (1/mk)[I_k - \mathbf{1}_k\mathbf{1}_k^{'}/k]$.

(b) Among all unbiased estimators of σ^2 in the above class, $\hat{\sigma}^2$ has the minimum variance if and only if $a_{rs} = 0$, for all $r \neq s$.

The proof of the theorem is given in the appendix at the end of this chapter.

Corollary 1 *The minimum variance unbiased nonnegative estimate of σ^2 in the class considered can be simplified as*

$$\hat{\sigma}^2_{UMVUE} = \frac{k(m-1)+1}{k^2m(m-1)}W^* + \frac{1}{mk}B^*,$$

where B^* and W^* are the between- and within-rank sum of squares of the RSS data defined by, respectively,

$$B^* = m\sum_{r=1}^{k}(\bar{X}_{[r]} - \bar{X}_{RSS})^2,$$
$$W^* = \sum_{r=1}^{k} W_r^*,$$

where $\bar{X}_{[r]} = (1/m)\sum_{i=1}^{m} X_{[r]i}$ and $W_r^* = \sum_{i=1}^{m}(X_{[r]i} - \bar{X}_{[r]})^2$, $r = 1, \cdots, k$. Moreover, we have

$$Var(\hat{\sigma}^2_{UMVUE}) = \frac{1}{mk^2}\sum_{r=1}^{k}\kappa^*_{[r]} + 2\frac{(k(m-1)+1)^2}{k^4m^2(m-1)}\sum_{r=1}^{k}\sigma^4_{[r]}$$
$$+\frac{2}{k^4m^2}[\sum_{r=1}^{k}\sigma^2_{[r]}]^2 + \frac{2k(k-2)}{k^4m^2}\sum_{r=1}^{k}\sigma^4_{[r]}$$
$$+\frac{4}{k^2m}\sum_{r=1}^{k}\tau^2_{[r]}\sigma^2_{[r]} + \frac{4}{k^2m}\sum_{r=1}^{k}\tau_{[r]}\gamma^*_{[r]},$$

where $\mu_{[r]}, \sigma_{[r]}, \gamma^*_{[r]}$ and $\kappa^*_{[r]}$ are, respectively the mean, standard deviation, third and fourth cumulants of $X_{[r]i}$, and $\tau_{[r]} = \mu_{[r]} - \mu$.

Remarks:

(i) It is easy to verify that two other unbiased nonnegative estimates of σ^2 with special choices of off-diagonal elements of A are given by

$$\hat{\sigma}_1^2 = \frac{k(m-1)+1}{km(m-1)}B + \frac{1}{mk}B^* \text{ and } \hat{\sigma}_2^2 = \frac{1}{mk}W + \frac{1}{k^2(m-1)}W^*.$$

(ii) The variance of Stokes's (1980) estimate of σ^2 can be easily obtained by choosing $A = \frac{1}{mk-1}\boldsymbol{I}_m$ and $D = \frac{1}{mk-1}[I_k - \mathbf{1}_k\mathbf{1}_k'/k]$.

(iii) The asymptotic relative efficiency (ARE) of $\hat{\sigma}^2_{\text{UMVUE}}$ compared to Stokes's estimate, defined as the limit of $\text{Var}(\hat{\sigma}^2_{Stokes})/\text{Var}(\hat{\sigma}^2_{\text{UMVUE}})$ as $m \to \infty$, turns out to be unity. However, for small m, the relative efficiency computed for several distributions shows the superiority of the UMVUE over Stokes's estimate. For details, see Perron and Sinha [132].

2.5 Tests and confidence intervals for population mean

In this section, we present asymptotic testing and confidence interval procedures for the population mean based on a balanced RSS under a nonparametric set up.

2.5.1 Asymptotic pivotal method

Assume that the population of concern has finite mean μ and variance σ^2. We discuss procedures for constructing confidence intervals and testing hypotheses for μ based on a pivot. A pivot for μ is a function of μ and the data, whose distribution or asymptotic distribution does not depend on any unknown parameters, and usually can be obtained by standardizing an unbiased estimate of μ. Thus if $\hat{\mu}_{\mathrm{RSS}}$ is an unbiased estimate of μ based on a balanced ranked set sample and $\hat{\sigma}_{\hat{\mu}_{\mathrm{RSS}}}$ is a consistent estimate of the standard deviation of $\hat{\mu}_{\mathrm{RSS}}$, a pivot for μ can be formed as

$$Z_1 = \frac{\hat{\mu}_{\mathrm{RSS}} - \mu}{\hat{\sigma}_{\hat{\mu}_{\mathrm{RSS}}}}.$$

Then an equal tailed $100(1-\alpha)\%$ confidence interval of μ can be constructed as

$$[\hat{\mu}_{\mathrm{RSS}} - z_{1-\alpha/2}\hat{\sigma}_{\hat{\mu}_{\mathrm{RSS}}}, \hat{\mu}_{\mathrm{RSS}} - z_{\alpha/2}\hat{\sigma}_{\hat{\mu}_{\mathrm{RSS}}}]$$

where $z_{\alpha/2}$ denotes the $(\alpha/2)$th quantile of the asymptotic distribution of Z_1. A hypothesis for μ, either one-sided or two-sided, can be tested based on the pivot Z_1 in a straightforward manner. On the other hand, based on a simple random sample of size N, a pivot Z_2 is given by

$$Z_2 = \frac{\bar{X} - \mu}{s/\sqrt{N}}$$

where $\bar{X}$ is the sample mean and s^2 is the usual unbiased estimate of σ^2. In what follows, we derive the counterparts of Z_2 by choosing Z_1 appropriately under a balanced ranked set sampling scheme.

Obviously, in the case of a *balanced* RSS, the ranked set sample mean, $\bar{X}_{\mathrm{RSS}} = 1/(mk)\sum_{r=1}^{k}\sum_{i=1}^{m} X_{[r]i}$, is an unbiased estimate of μ. Therefore, a pivot for μ can be formed as

$$Z = \frac{\bar{X}_{\mathrm{RSS}} - \mu}{\hat{\sigma}_{\bar{X}_{\mathrm{RSS}}}}$$

where $\hat{\sigma}_{\bar{X}_{\mathrm{RSS}}}$ is a consistent estimate of $\sigma_{\bar{X}_{\mathrm{RSS}}}$, the standard deviation of $\bar{X}_{\mathrm{RSS}}$. It follows from the central limit theorem that Z follows asymptotically a standard normal distribution. However, it turns out that there are several consistent estimates of $\sigma_{\bar{X}_{\mathrm{RSS}}}$, each of which gives rise to a pivotal statistic. The question then obviously arises as to which of the estimates of $\sigma_{\bar{X}_{\mathrm{RSS}}}$ should be used in the pivot. We shall now discuss different consistent estimates of $\sigma_{\bar{X}_{\mathrm{RSS}}}$ and make a choice in the next section.

2.5.2 Choice of consistent estimates of $\sigma_{\bar{X}_{\mathrm{RSS}}}$

In this section we discuss several consistent estimates of $\sigma_{\bar{X}_{\mathrm{RSS}}}$ and make a choice for the one to be used in the pivot considered in the previous section.

First, we give different expressions for $\bar{X}_{\text{RSS}}$ that motivate various estimates of $\sigma_{\bar{X}_{\text{RSS}}}$. We can write $\bar{X}_{\text{RSS}}$ as

$$\bar{X}_{\text{RSS}} = \frac{1}{m}\sum_{i=1}^{m}\frac{1}{k}\sum_{r=1}^{k} X_{[r]i} = \frac{1}{m}\sum_{i=1}^{m}\bar{X}_{i\text{RSS}}$$

where $\bar{X}_{i\text{RSS}}$, $i = 1, \cdots, m$, are i.i.d. with mean μ and variance, say, τ^2. Then we have

$$\text{Var}(\bar{X}_{\text{RSS}}) = \frac{\tau^2}{m}. \tag{2.11}$$

Obviously, an unbiased estimate of τ^2 is given by

$$\hat{\tau}_1^2 = \frac{1}{m-1}\sum_{i=1}^{m}(\bar{X}_{i\text{RSS}} - \bar{X}_{\text{RSS}})^2.$$

On the other hand, we can express $\bar{X}_{\text{RSS}}$ as

$$\bar{X}_{\text{RSS}} = \frac{1}{k}\sum_{r=1}^{k}\frac{1}{m}\sum_{i=1}^{m} X_{[r]i} = \frac{1}{k}\sum_{r=1}^{k}\bar{X}_{[r]}$$

where $\bar{X}_{[r]}$, $r = 1, \cdots, k$, are independent with means and variances given by, respectively, $E(\bar{X}_{[r]}) = \mu_{[r]}$ and $\text{Var}(\bar{X}_{[r]}) = \sigma_{[r]}^2/m$. Thus we have another expression of $\text{Var}(\bar{X}_{\text{RSS}})$ as follows.

$$\text{Var}(\bar{X}_{\text{RSS}}) = \frac{1}{mk^2}\sum_{r=1}^{k}\sigma_{[r]}^2. \tag{2.12}$$

Comparing (2.11) and (2.12), we have

$$\tau^2 = \frac{1}{k^2}\sum_{r=1}^{k}\sigma_{[r]}^2. \tag{2.13}$$

Hence an unbiased estimate of τ^2 is obtained through the expression (2.13) by forming unbiased estimates of $\sigma_{[r]}^2$'s. The usual unbiased estimate of $\sigma_{[r]}^2$ is given by

$$\hat{\sigma}_{[r]}^2 = \frac{1}{m-1}\sum_{i=1}^{m}(X_{[r]i} - \bar{X}_{[r]})^2,$$

since for fixed r, $X_{[r]i}, i = 1, \cdots, m$, are i.i.d. with variance $\sigma_{[r]}^2$. Hence, an unbiased estimate of τ^2 is given by

$$\hat{\tau}_2^2 = \frac{1}{k^2}\sum_{r=1}^{k}\hat{\sigma}_{[r]}^2.$$

Since $\hat{\sigma}^2_{[r]}$ is consistent for $\sigma^2_{[r]}$, obviously $\hat{\tau}_2^2$ is consistent for τ^2.

Finally, we can express τ^2 as

$$\tau^2 = \frac{\sigma^2}{k} - \frac{1}{k^2}\sum_{r=1}^{k}(\mu_{[r]} - \mu)^2. \tag{2.14}$$

Thus, if we plug in consistent estimates of σ^2, $\mu_{[r]}$ and μ into (2.14), we obtain a third consistent estimate of τ^2 as follows.

$$\hat{\tau}_3^2 = \frac{\hat{\sigma}^2}{k} - \frac{1}{k^2}\sum_{r=1}^{k}(\hat{\mu}_{[r]} - \hat{\mu})^2. \tag{2.15}$$

We can take

$$\begin{aligned}
\hat{\mu} &= \bar{X}_{\text{RSS}}, \\
\hat{\mu}_{[r]} &= \frac{1}{m}\sum_{i=1}^{m} X_{[r]i} = \bar{X}_{[r]}, \\
\hat{\sigma}^2 &= \frac{1}{mk}\sum_{r=1}^{k}\sum_{i=1}^{m}(X_{[r]i} - \bar{X}_{\text{RSS}})^2.
\end{aligned}$$

It can be verified that (see Stokes, 1980)

$$\begin{aligned}
E\hat{\sigma}^2 &= (1 - \frac{1}{mk})\sigma^2 + \frac{1}{mk^2}\sum_{r=1}^{k}(\mu_{[r]} - \mu)^2 \\
&= \sigma^2 + O(\frac{1}{m})
\end{aligned}$$

and that

$$\begin{aligned}
E[\frac{1}{k^2}\sum_{r=1}^{k}(\hat{\mu}_{[r]} - \hat{\mu})^2] &= \frac{1}{k^2}\sum_{r=1}^{k}(\mu_{[r]} - \mu)^2 + \frac{k-1}{mk}\tau^2 \\
&= \frac{1}{k^2}\sum_{r=1}^{k}(\mu_{[r]} - \mu)^2 + O(\frac{1}{m}).
\end{aligned}$$

Some other consistent estimates of τ^2 can also be obtained by plugging in bias-corrected estimators of σ^2 and $\sum_{r=1}^{k}(\mu_{[r]} - \mu)^2$ into (2.14).

In the remainder of this section, we compare the three estimates of τ^2 defined above. After some manipulation, we can obtain

$$\hat{\tau}_3^2 = \frac{m-1}{m}\hat{\tau}_2^2.$$

Therefore, $\hat{\tau}_3^2$ and $\hat{\tau}_2^2$ are asymptotically equivalent. In fact, any estimate of τ^2 obtained by bias-correcting $\hat{\sigma}^2$ and $\sum_{r=1}^{k}(\hat{\mu}_{[r]} - \hat{\mu})^2$ is asymptotically equivalent to $\hat{\tau}_2^2$. Such a bias-corrected estimator takes the form

$$a(m,k)\hat{\tau}_2^2 + b(m,k)\sum_{r=1}^{k}(\hat{\mu}_{[r]} - \hat{\mu})^2.$$

In order to make it consistent, we must have, as $m \to \infty$,

$$a(m,k) \to 1, \quad b(m,k) \to 0.$$

Hence, the estimate is essentially of the form

$$\hat{\tau}_2^2 + o(1).$$

The term $o(1)$ is of order $O(1/m^2)$ or higher with bias-correction.

Therefore, we only need to consider $\hat{\tau}_1^2$ and $\hat{\tau}_2^2$. Some straightforward algebra yields that

$$\hat{\tau}_1^2 = \hat{\tau}_2^2 + \frac{m}{(m-1)k^2}\sum_{r \neq s}(\bar{X}_{[rs]} - \bar{X}_{[r]}\bar{X}_{[s]}),$$

where

$$\bar{X}_{[rs]} = \frac{1}{m}\sum_{i=1}^{m} X_{[r]i}X_{[s]i}.$$

Note that

$$E\sum_{r \neq s}(\bar{X}_{[rs]} - \bar{X}_{[r]}\bar{X}_{[s]}) = 0.$$

We have the following result.

Lemma 2.5. *Among all unbiased estimates of τ^2 of the form*

$$\hat{\tau}_2^2 + \lambda\sum_{r \neq s}(\bar{X}_{[rs]} - \bar{X}_{[r]}\bar{X}_{[s]}),$$

the estimate $\hat{\tau}_2^2$ has the smallest variance.

Proof: In order to prove the lemma, we only need to verify that, for any l, r, s such that $r \neq s$,

$$\text{Cov}(\bar{X}_{[ll]} - \bar{X}_{[l]}^2, \bar{X}_{[rs]} - \bar{X}_{[r]}\bar{X}_{[s]}) = 0. \tag{2.16}$$

If both r and s are not equal to l, (2.16) holds trivially since $X_{[l]i}$'s are independent from $X_{[r]i}$'s and $X_{[s]i}$'s. In the following, we verify (2.16) for the case that either r or s equals l. Without loss of generality, let $r = l$. We have

$$\begin{aligned}
&\text{Cov}(\bar{X}_{[ll]} - \bar{X}_{[l]}^2, \bar{X}_{[ls]} - \bar{X}_{[l]}\bar{X}_{[s]}) \\
&= E[\text{Cov}(\bar{X}_{[ll]} - \bar{X}_{[l]}^2, \bar{X}_{[ls]} - \bar{X}_{[l]}\bar{X}_{[s]} | X_{[s]i}, i = 1, \ldots, m)] \\
&= E[\frac{1}{m}\sum_{i=1}^{m} X_{[s]i}\text{Cov}(\bar{X}_{[ll]} - \bar{X}_{[l]}^2, X_{[l]i})] - E[\bar{X}_{[s]}\text{Cov}(\bar{X}_{[ll]} - \bar{X}_{[l]}^2, \bar{X}_{[l]})] \\
&= 0.
\end{aligned}$$

As a special case of Lemma 2.5, we have

$$\mathrm{Var}(\hat{\tau}_2^2) < \mathrm{Var}(\hat{\tau}_1^2).$$

We summarize what we have found so far to conclude this section. We have investigated possible ways of estimating the variance of a standard RSS estimate of the population mean. We compared different estimates of this variance and found the one with the smallest variance. We recommend that the estimate with the smallest variance be used in the pivot for the construction of confidence intervals or testing hypotheses for the population mean.

Thus, the pivot recommended is

$$Z = \frac{\sqrt{mk}(\bar{X}_{\text{RSS}} - \mu)}{\hat{\tau}_2}.$$

2.5.3 Comparison of power between tests based on RSS and SRS

In this section, we make a power comparison between the test using pivot Z based on RSS and the test using pivot Z_2 based on SRS by a simulation study. In the simulation study, the values of k and m are taken as $k = 3, 4, 5$, and $m = 15, 20, 25$. The underlying distribution is taken as normal distributions $N(\mu, 1)$ with a number of μ values. The power of the SRS-based test is computed by using the standard normal distribution since the sample size $N = mk$ is large. The power of the RSS based test is simulated. In Table 2.3 below, we report the power of the two tests for testing $H_0 : \mu = 0$ against $H_1 : \mu > 0$ at the significance level $\alpha = 0.05$. The efficiency of RSS over SRS is obvious from Table 2.3.

2.6 Estimation of quantiles

This section is devoted to the estimation of population quantiles by using a balanced ranked set sample. We first give the definition of the ranked set sample quantiles analogous to the simple random sample quantiles and investigate their properties. Then we consider inference procedures such as the construction of confidence intervals and the testing of hypotheses for the quantiles. Finally we make a comparison between RSS quantile estimates and SRS quantile estimates in terms of their asymptotic variances.

2.6.1 Ranked set sample quantiles and their properties

Let the ranked-set empirical distribution function be defined as

$$\hat{F}_{\text{RSS}}(x) = \frac{1}{mk} \sum_{r=1}^{k} \sum_{i=1}^{m} I\{X_{[r]i} \leq x\}.$$

Table 2.3. Power of tests based on Z and Z_2

m	k	Pivot\ μ	0	.01	.02	.05	.1	.2	.3	.4	.5
15	3	Z	.05	.07	.08	.12	.25	.59	.97	1	1
		Z_2	.05	.06	.06	.09	.16	.38	.85	.96	1
	4	Z	.06	.09	.13	.19	.40	.83	1	1	1
		Z_2	.05	.06	.07	.10	.19	.46	.93	.99	1
	5	Z	.06	.07	.09	.19	.46	.90	1	1	1
		Z_2	.05	.06	.07	.11	.22	.53	.96	1	1
20	3	Z	.05	.06	.07	.12	.29	.64	.99	1	1
		Z_2	.05	.06	.07	.10	.19	.46	.93	.99	1
	4	Z	.06	.09	.12	.22	.43	.90	1	1	1
		Z_2	.05	.06	.07	.12	.23	.56	.97	1	1
	5	Z	.05	.08	.09	.22	.51	.94	1	1	1
		Z_2	.05	.06	.07	.13	.26	.64	.99	1	1
25	3	Z	.05	.06	.07	.16	.33	.78	1	1	1
		Z_2	.05	.06	.07	.11	.22	.53	.97	.99	1
	4	Z	.06	.07	.09	.12	.46	.92	1	1	1
		Z_2	.05	.06	.07	.13	.26	.72	1	1	1
	5	Z	.04	.06	.11	.22	.58	.98	1	1	1
		Z_2	.05	.06	.08	.14	.30	.72	1	1	1

Let $n = mk$. For $0 < p < 1$, the pth ranked-set sample quantile, denoted by $\hat{x}_n(p)$, is then defined as the pth quantile of $\hat{F}_{\text{RSS}}$, i.e.,

$$\hat{x}_n(p) = \inf\{x : \hat{F}_{\text{RSS}}(x) \geq p\}.$$

We also define the ranked-set order statistics as follows. Let the $X_{[r]i}$'s be ordered from the smallest to the largest and denote the ordered quantities by

$$Z_{(1:n)} \leq \cdots \leq Z_{(j:n)} \leq \cdots \leq Z_{(n:n)}.$$

The $Z_{(j:n)}$'s are then referred to as the ranked-set order statistics.

The following results are parallel to those on simple random sample quantiles.

Let the pth quantile of F be denoted by $x(p)$. First, we have that $\hat{x}_n(p)$ converges with probability one to $x(p)$. Indeed, more strongly, we have

Theorem 2.6. *Suppose that the ranking mechanism in RSS is consistent. Then, with probability 1,*

$$|\hat{x}_n(p) - x(p)| \leq \frac{2(\log n)^2}{f(x(p))n^{1/2}},$$

for all sufficiently large n.

Next, we have a Bahadur representation for the ranked-set sample quantile.

Theorem 2.7. *Suppose that the ranking mechanism in RSS is consistent and that the density function f is continuous at $x(p)$ and positive in a neighborhood of $x(p)$. Then,*

$$\hat{x}_n(p) = x(p) + \frac{p - \hat{F}_{RSS}(x(p))}{f(x(p))} + R_n,$$

where, with probability one,

$$R_n = O(n^{-3/4}(\log n)^{3/4}),$$

as $n \to \infty$.

From the Bahadur representation follows immediately the asymptotic normality of the ranked-set sample quantile.

Theorem 2.8. *Suppose that the same conditions as in Theorem 2.7 hold. Then*

$$\sqrt{n}(\hat{x}_n(p) - x(p)) \to N\left(0, \frac{\sigma_{k,p}^2}{f^2(x(p))}\right),$$

in distribution, where,

$$\sigma_{k,p}^2 = \frac{1}{k}\sum_{r=1}^{k} F_{[r]}(x(p))[1 - F_{[r]}(x(p))].$$

In particular, if ranking is perfect, noting that $F_{[r]}(x(p)) = B(r, k+r-1, p)$ in this case,

$$\sigma_{k,p}^2 = \frac{1}{k}\sum_{r=1}^{k} B(r, k+r-1, p)[1 - B(r, k+r-1, p)],$$

where $B(r, s, x)$ denotes the distribution function of the beta distribution with parameters r and s.

Notice that, when ranking is perfect, the quantity $\sigma_{k,p}^2$ does not depend on any unknowns, which is practically important as will be seen when the asymptotic normality is applied to develop inference procedures for the population quantile. In general, $F_{[r]}(x(p))$ depends on both the ranking mechanism and the unknown F, and needs to be estimated from the data.

As another immediate consequence of Theorem 2.7, we have a more general result as follows.

Theorem 2.9. *Let $0 < p_1 < \cdots < p_j \cdots < p_l < 1$ be l probabilities. Let $\boldsymbol{\xi} = (x(p_1), \ldots, x(p_l))^T$ and $\hat{\boldsymbol{\xi}} = (\hat{x}_n(p_1), \ldots, \hat{x}_n(p_l))^T$. Then*

$$\sqrt{n}(\hat{\boldsymbol{\xi}} - \boldsymbol{\xi}) \to N_l(\mathbf{0}, \Sigma)$$

in distribution where, for $i \le j$, the (i,j)-th entry of Σ is given by

$$\sigma_{ij} = \frac{1}{k}\sum_{r=1}^{k} F_{[r]}(x(p_i))(1 - F_{[r]}(x(p_j)))/[f(x(p_i))f(x(p_j))].$$

We also derive some properties of the ranked-set order statistics. An order statistic $Z_{(k_n:n)}$ is said to be central if $\frac{k_n}{n}$ converges to some p such that $0 < p < 1$ as n goes to infinity. For central ranked-set order statistics, we have the following analogue of the results for simple random sample order statistics:

Theorem 2.10. *(i) If* $\frac{k_n}{n} = p + o(n^{-1/2})$ *then*

$$Z_{(k_n:n)} = x(p) + \frac{\frac{k_n}{n} - \hat{F}_{RSS}(x(p))}{f(x(p))} + R_n,$$

where, with probability 1,

$$R_n = O(n^{-3/4}(\log n)^{3/4}),$$

as $n \to \infty$.

(ii) If

$$\frac{k_n}{n} = p + \frac{c}{n^{1/2}} + o(n^{-1/2})$$

then

$$\sqrt{n}(Z_{(k_n:n)} - \hat{x}_n(p)) \to \frac{c}{f(x(p))}$$

with probability 1, and

$$\sqrt{n}(Z_{(k_n:n)} - x(p)) \to N\left(\frac{c}{f(x(p))}, \frac{\sigma_{k,p}^2}{f^2(x(p))}\right)$$

in distribution.

The results in this section can be proved in arguments parallel to the proof of the results on simple random sample quantiles. The details of the proof can be found in Chen [35].

2.6.2 Inference procedures for population quantiles based on ranked set sample

The results in Section 2.6.1 are applied in this section for inference procedures on quantiles such as confidence intervals and hypotheses testing.

(i) Confidence interval based on ranked-set order statistics. To construct a confidence interval of confidence coefficient $1 - 2\alpha$ for $x(p)$, we seek two integers l_1 and l_2 such that $1 \le l_1 < l_2 \le n$ and that

$$P(Z_{(l_1:n)} < x(p) < Z_{(l_2:n)}) = 1 - 2\alpha.$$

We restrict our attention to the intervals with equal tail probabilities, i.e., intervals satisfying

$$P(Z_{(l_1:n)} \leq x(p)) = 1 - \alpha, \quad P(Z_{(l_2:n)} \leq x(p)) = \alpha.$$

Then the integers l_1 and l_2 can be found as follows. Let N_r denote the number of $X_{[r]i}$'s with fixed r which are less than or equal to $x(p)$. Let $N = \sum_{r=1}^{k} N_r$. We have

$$P(Z_{(l_1:n)} \leq x(p)) = P(N \geq l_1).$$

Note that the N_r's are independent binomial random variables with $N_r \sim Bi(m, p_r)$ where $p_r = F_{[r]}(x(p))$. Hence

$$P(N \geq l_1) = \sum_{j=l_1}^{n} \sum_{(j)} \prod_{r=1}^{k} \binom{m}{i_r} p_r^{i_r} (1 - p_r)^{m - i_r},$$

where the summation $\sum_{(j)}$ is over all k-tuples of integers $(i_1, \ldots, i_k)$ satisfying $\sum_{r=1}^{k} i_r = j$. Then l_1 can be determined such that the sum on the right hand of the above equality is equal to or near $1 - \alpha$. Similarly l_2 can be determined. Though not impossible, the computation will be extremely cumbersome. However, when n is large, l_1 and l_2 can be determined approximately as demonstrated below. Note that

$$EN = \sum_{r=1}^{k} mF_{[r]}(x(p)) = mkF(x(p)) = np,$$

$$Var(N) = \sum_{r=1}^{k} mF_{[r]}(x(p))[1 - F_{[r]}(x(p))].$$

By the central limit theorem we have that, approximately,

$$\frac{N - np}{\sqrt{\sum_{r=1}^{k} mF_{[r]}(x(p))[1 - F_{[r]}(x(p))]}} \sim N(0, 1).$$

Hence

$$l_1 \approx np - z_\alpha \sqrt{\sum_{r=1}^{k} mF_{[r]}(x(p))[1 - F_{[r]}(x(p))]},$$

$$l_2 \approx np + z_\alpha \sqrt{\sum_{r=1}^{k} mF_{[r]}(x(p))[1 - F_{[r]}(x(p))]},$$

where z_α denotes the $(1 - \alpha)$th quantile of the standard normal distribution. When ranking is perfect, $F_{[r]}(x(p)) = B(r, k-r+1, p)$, and the intervals above can be completely determined. However, in general, $F_{[r]}(x(p))$ is unknown and has to be estimated. We can take the estimate to be $\hat{F}_{[r]}(\hat{x}_n(p))$, where $\hat{F}_{[r]}(x) = (1/m) \sum_{i=1}^{m} I\{X_{[r]i} \leq x\}$.

For later reference, the interval $[Z_{(l_1:n)}, Z_{(l_2:n)}]$ is denoted by $\tilde{I}_{S_n}$.

(ii) Confidence interval based on ranked-set sample quantiles. By making use of Theorem 2.8, another asymptotic confidence interval of confidence coefficient $1-2\alpha$ for $x(p)$ based on ranked-set sample quantiles can be constructed as:

$$\left[\hat{x}_n(p) - \frac{z_\alpha}{\sqrt{n}} \frac{\sigma_{k,p}}{f(x(p))}, \ \hat{x}_n(p) + \frac{z_\alpha}{\sqrt{n}} \frac{\sigma_{k,p}}{f(x(p))}\right].$$

This interval is denoted by $\tilde{I}_{Q_n}$. Since $\tilde{I}_{Q_n}$ involves the unknown quantity $f(x(p))$, we need to replace it with some consistent estimate in practice. In the next section, we shall consider the estimation of f by the kernel method using RSS data. The RSS kernel estimate of f can well serve the purpose here. Let $\hat{f}_{\text{RSS}}$ denote the RSS kernel estimate of f. Then in $\tilde{I}_{Q_n}$ the unknown $f(x(p))$ can be replaced by $\hat{f}_{\text{RSS}}(\hat{x}_n(p))$. Note that the intervals $\tilde{I}_{S_n}$ and $\tilde{I}_{Q_n}$ are equivalent in the sense that the two intervals are approximately overlapping with each other while the confidence coefficients are the same. It follows from Theorems 2.7 and 2.10 that, with probability 1,

$$Z_{(l_1:n)} - [\hat{x}_n(p) - \frac{z_\alpha}{\sqrt{n}} \frac{\sigma_{k,p}}{f(x(p))}] = o(n^{-1/2}),$$

and

$$Z_{(l_2:n)} - [\hat{x}_n(p) + \frac{z_\alpha}{\sqrt{n}} \frac{\sigma_{k,p}}{f(x(p))}] = o(n^{-1/2}).$$

Noting that the length of the two intervals has order $O(n^{-1/2})$, the equivalence is established. In practice, either of these two intervals could be used.

(iii) Hypothesis testing using ranked-set sample quantiles. The joint asymptotic normality of the ranked-set sample quantiles, as stated in Theorem 2.9, can be used to test hypotheses involving population quantiles. Suppose the null hypothesis is of the form $\boldsymbol{l}^T\boldsymbol{\xi} = c$, where $\boldsymbol{\xi}$ is a vector of quantiles, say, $\boldsymbol{\xi} = (x(p_1), \ldots, x(p_q))^T$, and $\boldsymbol{l}$ and c are given vector and scalar of constants, respectively. The test statistic can then be formed as:

$$S_n = \frac{\sqrt{n}[\boldsymbol{l}^T\hat{\boldsymbol{\xi}} - c]}{\sqrt{\boldsymbol{l}^T\hat{\Sigma}\boldsymbol{l}}},$$

where $\hat{\boldsymbol{\xi}}$ is the vector of the corresponding ranked-set sample quantiles and $\hat{\Sigma}$ is the estimated covariance matrix of $\hat{\boldsymbol{\xi}}$ with its (i,j)-th $(i < j)$ entry given by

$$\hat{\sigma}_{ij} = \frac{1}{k}\sum_{r=1}^{k} p_{ir}(1-p_{jr})/[\hat{f}(\hat{x}_n(p_i))\hat{f}(\hat{x}_n(p_j))].$$

By Theorem 2.9, the test statistic follows asymptotically the standard normal distribution under the null hypothesis. Hence, the decision rule can be made accordingly.

2.6.3 The relative efficiency of RSS quantile estimate with respect to SRS quantile estimate

We discuss the ARE of the RSS quantile estimate with respect to the SRS quantile estimate in this section. The counterpart of $\hat{x}_n(p)$ in SRS is the pth sample quantile, $\hat{\xi}_{np}$, of a simple random sample of size n. It can be found from any standard text book that $\hat{\xi}_{np}$ has an asymptotic normal distribution with mean $x(p)$ and variance $\frac{p(1-p)}{nf^2(x(p))}$. Hence, the ARE of $\hat{x}_n(p)$ with respect to $\hat{\xi}_{np}$ is given by

$$\text{ARE}(\hat{x}_n(p), \hat{\xi}_{np}) = \frac{p(1-p)}{\frac{1}{k}\sum_{r=1}^{k} F_{[r]}(x(p))[1 - F_{[r]}(x(p))]}.$$

By using the Bahadur representations of $\hat{x}_n(p)$ and $\hat{\xi}_{np}$ and applying Theorem 2.1 to the function

$$h(x) = \frac{p - I\{x \leq x(p)\}}{f(x(p))},$$

we obtain that

$$\text{ARE}(\hat{x}_n(p), \hat{\xi}_{np}) > 1,$$

provided that the ranking mechanism in RSS is consistent.

While $\text{ARE}(\hat{x}_n(p), \hat{\xi}_{np})$ is always greater than 1 for any p, the quantity can differ very much for different values of p. To gain more insight into the nature of the ARE, let us consider the case of perfect ranking. In this case, $F_{[r]}(x(p)) = B(r, k-r+1, p)$, and the relative efficiency depends only on p and k. For convenience, let it be denoted by $\text{ARE}(k,p)$. For fixed k, as a function of p, $\text{ARE}(k,p)$ is symmetric about $p = 0.5$. It achieves its maximum at $p = 0.5$ and damps away towards $p = 0$ and $p = 1$. For fixed p, $\text{ARE}(k,p)$ increases as k increases. For $k = 1, \ldots, 10$, $ARE(k,p)$ is depicted in Figure 2.1. The curves from the bottom to the top correspond to k from 1 to 10.

We can expect the largest gain in efficiency when we estimate the median of a population. The gain is quite significant even for small set sizes. For $k = 3, 4, 5$, the AREs are, respectively, 1.6, 1.83, 2.03 — but relatively poor compared with the RE's for mean for most distributions — see Table 2.1. In terms of sample sizes, we can reduce the sample size of an SRS by a half through RSS with set size $k = 5$ while maintaining the same accuracy. However, the efficiency gain in the estimation of extreme quantiles is almost negligible. To improve the efficiency for the estimation of extreme quantiles, other RSS procedures must be sought. We will get back to this point in Chapter 4.

2.7 Estimation of density function with ranked set sample

In the context of RSS, the need for density estimation arises in certain statistical procedures. For example, the confidence interval and hypothesis testing

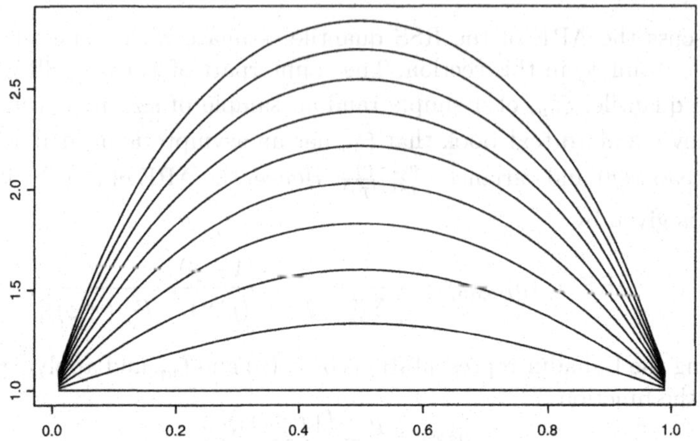

Fig. 2.1. The asymptotic relative efficiency of RSS quantile estimates

procedures based on ranked-set sample quantiles considered in Section 2.6 need to have an estimate of the values of the density function at certain quantiles. On the other hand, density estimation has its own independent interest. A density estimate can reveal important features such as the skewness and multimodality of the underlying distribution. A density estimate is an ideal tool for the presentation of the data back to the clients in order to provide explanations and illustrations of the conclusions that have been obtained. In this section, we take up the task for developing methods of density estimation using RSS data. In Section 2.7.1, the estimate of the density function f is given and its properties are investigated. In Section 2.7.2, the relative efficiency of the density estimate using RSS data with respect to its counterpart in SRS is discussed.

2.7.1 RSS density estimate and its properties

There is a vast literature on density estimation in SRS. A variety of methods have been proposed and developed including the nearest neighbor, the kernel, the maximum penalized likelihood and the adaptive kernel method, etc.. A good reference on the general methodology of density estimation is Silverman [153]. Each of the various methods has its own merits and drawbacks. There is no universal agreement as to which method should be used. We will focus our attention on the kernel method of density estimation and its ramifications. We choose to deal with the kernel method partly because it is a good choice in many practical problems and partly because its mathematical properties are well understood.

To facilitate our discussion that follows, before we deal with the properties of the RSS estimate, we give the definition of the SRS estimate along with some of its properties below. Based on a simple random sample $X_1, \ldots, X_n$, the kernel estimate of f is given by

$$\hat{f}_{\text{SRS}}(x) = \frac{1}{nh} \sum_{i=1}^{n} K\left(\frac{x - X_i}{h}\right).$$

The mean and variance of $\hat{f}_{\text{SRS}}(x)$ can be easily derived as

$$E\hat{f}_{\text{SRS}}(x) = \int \frac{1}{h} K\left(\frac{x-t}{h}\right) f(t)dt, \tag{2.17}$$

$$Var\hat{f}_{\text{SRS}}(x) = \frac{1}{n}\left\{\int \frac{1}{h^2} K\left(\frac{x-t}{h}\right)^2 f(t)dt - \left[\int \frac{1}{h} K\left(\frac{x-t}{h}\right) f(t)dt\right]^2\right\}. \tag{2.18}$$

To motivate the definition of the estimate with RSS data, note that, from the fundamental equality, we have

$$f = \frac{1}{k} \sum_{r=1}^{k} f_{[r]}, \tag{2.19}$$

where $f_{[r]}$ denotes the density function corresponding to $F_{[r]}$. The sub-sample, $X_{[r]i}, i = 1, \ldots, m$, is indeed a simple random sample from the distribution with pdf $f_{[r]}$. Hence $f_{[r]}$ can be estimated by the usual kernel method using the sub-sample. The kernel estimate, $\hat{f}_{[r]}$, of $f_{[r]}$ at x based on the sub-sample is defined as

$$\hat{f}_{[r]}(x) = \frac{1}{mh} \sum_{i=1}^{m} K\left(\frac{x - X_{[r]i}}{h}\right),$$

where K is a kernel function and h is the bandwidth to be determined. Thus a natural definition of the kernel estimate of f is given by

$$\hat{f}_{\text{RSS}}(x) = \frac{1}{k} \sum_{r=1}^{k} \hat{f}_{[r]}(x) = \frac{1}{kmh} \sum_{r=1}^{k} \sum_{i=1}^{m} K\left(\frac{x - X_{[r]i}}{h}\right).$$

It follows from (2.19) that

$$E\hat{f}_{\text{RSS}}(x) = \frac{1}{kh} \sum_{r=1}^{k} EK\left(\frac{x - X_{[r]i}}{h}\right)$$

$$= \frac{1}{k}\sum_{r=1}^{k}\int K\left(\frac{x-t}{h}\right) f_{[r]}(t)dt$$
$$= \int \frac{1}{h}K\left(\frac{x-t}{h}\right) f(t)dt = E\hat{f}_{\mathrm{SRS}}(x),$$

$$Var\hat{f}_{\mathrm{RSS}}(x) = \frac{1}{mk^2}\sum_{r=1}^{k} Var\frac{1}{h}K\left(\frac{x-X_{[r]}}{h}\right)$$
$$= \frac{1}{mk^2}\sum_{r=1}^{k}\left\{E\left[\frac{1}{h}K\left(\frac{x-X_{[r]}}{h}\right)\right]^2 - \left[E\frac{1}{h}K\left(\frac{x-X_{[r]}}{h}\right)\right]^2\right\}$$
$$= \frac{1}{mk}\left\{E\left[\frac{1}{h}K\left(\frac{x-X}{h}\right)\right]^2 - \frac{1}{k}\sum_{r=1}^{k}\left[E\frac{1}{h}K\left(\frac{x-X_{[r]}}{h}\right)\right]^2\right\}$$
$$= \mathrm{Var}\hat{f}_{\mathrm{SRS}}(x)$$
$$+ \frac{1}{mk}\left\{\left[E\frac{1}{h}K\left(\frac{x-X}{h}\right)\right]^2 - \frac{1}{k}\sum_{r=1}^{k}\left[E\frac{1}{h}K\left(\frac{x-X_{[r]}}{h}\right)\right]^2\right\}.$$

It follows again from (2.19) that

$$E\frac{1}{h}K\left(\frac{x-X}{h}\right) = \frac{1}{k}\sum_{r=1}^{k} E\frac{1}{h}K\left(\frac{x-X_{[r]}}{h}\right).$$

By the Cauchy-Schwarz inequality we have

$$\left[E\frac{1}{h}K\left(\frac{x-X}{h}\right)\right]^2 < \frac{1}{k}\sum_{r=1}^{k}\left[E\frac{1}{h}K\left(\frac{x-X_{[r]}}{h}\right)\right]^2.$$

Summarizing the argument above we conclude that $\hat{f}_{\mathrm{RSS}}(x)$ has the same expectation as $\hat{f}_{\mathrm{SRS}}(x)$ and a smaller variance than $\hat{f}_{\mathrm{SRS}}(x)$. This implies that the RSS estimate has a smaller mean integrated square error (MISE) than the SRS estimate. The MISE of an estimate $\hat{f}$ of f is defined as $\mathrm{MISE}(\hat{f}) = E\int[\hat{f}(x) - f(x)]^2dx$. The conclusion holds whether or not ranking is perfect. In what follows we assume that f has certain derivatives and that K satisfies the conditions: (i) K is symmetric and (ii) $\int K(t)dt = 1$ and $\int t^2K(t)dt \neq 0$.

Lemma 2.11. *Under the above assumptions about f and K, for fixed k, as $h \to 0$,*

$$\left[E\frac{1}{h}K\left(\frac{x-X}{h}\right)\right]^2 - \frac{1}{k}\sum_{r=1}^{k}\left[E\frac{1}{h}K\left(\frac{x-X_{[r]}}{h}\right)\right]^2$$
$$= [f^2(x) - \frac{1}{k}\sum_{r=1}^{k} f_{[r]}^2(x)] + O(h^2).$$

Lemma 2.11 can be proved by a straightforward calculation involving Taylor expansions of f and $f_{[r]}$'s.

Let

$$\Delta(f,k) = \int [\frac{1}{k}\sum_{r=1}^{k} f_{[r]}^2(x) - f^2(x)]dx.$$

Note that $\Delta(f,k)$ is always greater than zero. We have the following result.

Theorem 2.12. *Suppose that the same bandwidth is used in both $\hat{f}_{SRS}$ and $\hat{f}_{RSS}$. Then, for fixed k and large n,*

$$MISE(\hat{f}_{RSS}) = MISE(\hat{f}_{SRS}) - \frac{1}{n}\Delta(f,k) + O\left(\frac{h^2}{n}\right).$$

We now consider the special case of perfect ranking and derive some asymptotic results that shed lights on the properties of the RSS density estimate. When ranking is perfect, $f_{[r]} = f_{(r)}$, the pdf of the rth order statistic. First, we have

Lemma 2.13. *If, for $r = 1, \ldots, k$, $f_{(r)}$ is the density function of the r-th order statistic of a sample of size k from a distribution with density function f, then we have the representation*

$$\frac{1}{k}\sum_{r=1}^{k} f_{(r)}^2(x) = kf^2(x)P(Y = Z),$$

where Y and Z are independent with the same binomial distribution $B(k-1, F(x))$. Furthermore

$$P(Y = Z) = \frac{1}{\sqrt{4\pi kF(x)[1 - F(x)]}} + o\left(\frac{1}{k}\right).$$

Proof: When ranking is perfect, we have

$$f_{(r)}(x) = \frac{k!}{(r-1)!(k-r)!}F^{r-1}(x)[1 - F(x)]^{k-r}f(x).$$

Thus, we can write

$$\begin{aligned}
&\frac{1}{k}\sum_{r=1}^{k} f_{(r)}^2(x) \\
&= \frac{1}{k}\sum_{r=1}^{k}[\frac{k!}{(r-1)!(k-r)!}F^{r-1}(x)(1 - F(x))^{k-r}f(x)]^2 \\
&= kf^2(x)\sum_{j=0}^{k-1}\left[\binom{k-1}{j}F^j(x)(1 - F(x))^{k-1-j}\right]^2.
\end{aligned}$$

The first part of the lemma is proved. The second part follows from the Edgeworth expansion of the probability $P(Y = Z)$.

Remark: Our computation for certain values of $F(x)$ has revealed that the approximation to the probability $P(Y = Z)$ is quite accurate for large or moderate k. For small k, $P(Y = Z)$ is slightly bigger than the approximation. However, the approximation can well serve our theoretical purpose.

In what follows we denote, for any function g, the integral $\int x^l g(x)dx$ by $i_l(g)$. Applying Lemmas 2.11 and 2.13, we have

Lemma 2.14. *If ranking is perfect, then, for a fixed large or moderate k, as $n \to \infty$, we have*

$$MISE(\hat{f}_{RSS}) = MISE(\hat{f}_{SRS}) - \frac{1}{n}[\sqrt{k}\delta(f) - i_0(f^2)] + O(\frac{h^2}{n}),$$

where

$$\delta(f) = \frac{1}{2\sqrt{\pi}} \int \frac{f^2(x)}{\sqrt{F(x)[1 - F(x)]}} dx.$$

Lemma 2.14 shows that the RSS estimate reduces the MISE of the SRS estimate at the order $O(n^{-1})$ by an amount which increases linearly in $\sqrt{k}$.

The results derived in this section can be extended straightforwardly to the adaptive kernel estimation described in Silverman ([153], p101). The ordinary kernel estimate usually suffers a slight drawback that it has a tendency of undersmoothing at the tails of the distribution. The adaptive kernel estimate overcomes this drawback and provides better estimates at the tails. We do not discuss the adaptive kernel estimate further. The reader is referred to Silverman ([153], Chapter 5) for details.

2.7.2 The relative efficiency of the RSS density estimate with respect to its SRS counterpart

In this section, we investigate the efficiency of the RSS estimate relative to the SRS estimate in terms of the ratio of the MISE's.

First, we derive an asymptotic expansion for the MISE of the SRS estimate. By Taylor expansion of the density function f at x under the integrals in (2.17) and (2.18) after making the change of variable $y = (x - t)/h$, we have

$$\text{bias}(\hat{f}_{\text{SRS}}(x)) = \frac{1}{2} i_2(K) f''(x) h^2 + O(h^4),$$

$$\text{Var}(\hat{f}_{\text{SRS}}(x)) = \frac{1}{nh} i_0(K^2) f(x) - \frac{1}{n} f^2(x) + O(\frac{h^2}{n}).$$

Hence

$$\text{MISE}(\hat{f}_{\text{SRS}}) = \int [Var(\hat{f}_{\text{SRS}}(x)) + bias^2(\hat{f}_{\text{SRS}}(x))] dx$$

$$= \frac{1}{nh} i_0(K^2) + \frac{1}{4} i_2^2(K) i_0(f''^2) h^4 - \frac{1}{n} i_0(f^2) + O(\frac{h^2}{n}) + O(h^6). \tag{2.20}$$

Minimizing the leading terms with respect to h, we have that the minimum is attained at

$$h_{opt} = i_2(K)^{-2/5} \left[\frac{i_0(K^2)}{i_0(f''^2)} \right]^{1/5} n^{-1/5}. \tag{2.21}$$

Substituting (2.21) into (2.20) yields

$$\text{MISE}(\hat{f}_{\text{SRS}}) = \frac{5}{4} C(K) i_0(f''^2)^{1/5} n^{-4/5} - i_0(f^2) n^{-1} + O(n^{-6/5}), \tag{2.22}$$

where $C(K) = i_2(K)^{2/5} i_0(K^2)^{4/5}$.

Combining (2.22) with Theorem 2.12, we have that the relative efficiency of the RSS estimate to the SRS estimate is approximated by

$$\frac{\text{MISE}(\hat{f}_{\text{SRS}})}{\text{MISE}(\hat{f}_{\text{RSS}})} \approx \left[1 - \frac{\Delta(f,k)}{(5/4)C(K) i_0(f''^2) n^{1/5} - i_0(f^2)} \right]^{-1}. \tag{2.23}$$

When ranking is perfect and k is large or moderate, the relative efficiency has the approximate expression:

$$\frac{\text{MISE}(\hat{f}_{\text{SRS}})}{\text{MISE}(\hat{f}_{\text{RSS}})} \approx \left[1 - \frac{\sqrt{k}\delta(f) - i_0(f^2)}{(5/4)C(K) i_0(f''^2) n^{1/5} - i_0(f^2)} \right]^{-1}. \tag{2.24}$$

We can conclude qualitatively from the approximation in (2.24) that (i) the efficiency of the RSS kernel estimate relative to the SRS kernel estimate increases as k increases at the rate $O(k^{1/2})$, (ii) the relative efficiency damps away as n gets large but the speed at which it damps away is very low (of order $O(n^{-1/5})$). Therefore, we can expect that for small or moderate sample size n the gain in efficiency by using RSS will be substantial. RSS can only reduce variance and the order $O(n^{-1})$ at which the variance is reduced is common in all the other statistical procedures such as the estimation of mean, variance and cumulative distribution, etc.. However, while the reduction in MISE is at order $O(n^{-1})$, the MISEs have order $O(n^{-4/5})$. When n is large, the component of the MISE at order $O(n^{-4/5})$ dominates. This explains the fact that the relative efficiency damps away as n goes to infinity.

A major application of the RSS density estimation is for estimating the density at certain particular points, e.g., certain quantiles. It is desirable to compare the performance of the RSS estimate and the SRS estimate at particular values of x. An argument similar to the global comparison leads to the following results. The MSEs of the two estimates at x have the equal components at order lower than $O(n^{-1})$. The components of the MSEs at order $O(n^{-1})$

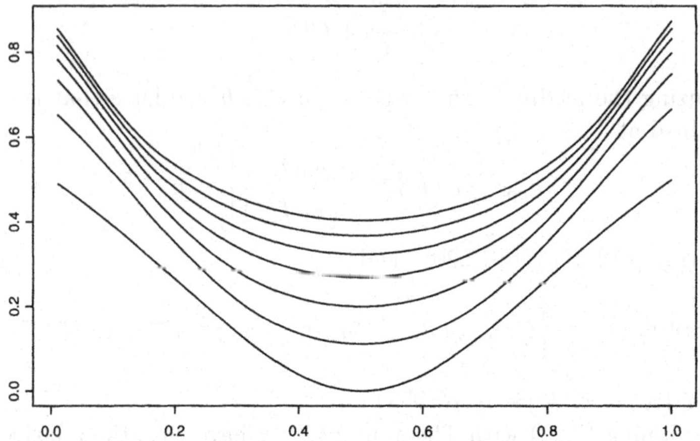

Fig. 2.2. The relative reduction in variance at order $O(n^{-1})$ of the RSS density estimate

of the SRS estimate and the RSS estimate are, respectively, $-(1/n)f^2(x)$ and $-(1/nk)\sum_{r=1}^{k} f_{[r]}^2(x)$. The relative reduction in MSE is then given by

$$\Lambda_f = [\frac{1}{k}\sum_{r=1}^{k} f_{[r]}^2(x) - f^2(x)]/\frac{1}{k}\sum_{r=1}^{k} f_{[r]}^2(x).$$

If ranking is perfect and $x = x(p)$, the pth quantile of the underlying distribution, then Λ_f becomes

$$\Lambda_f(p) = 1 - \left\{k\sum_{r=0}^{k-1}\left[\binom{k-1}{r} p^r(1-p)^{k-1-r}\right]^2\right\}^{-1}.$$

For $k = 2, \ldots, 8$, the relative reduction $\Lambda_f(p)$ is plotted against p in Figure 2.2. The curves from bottom to top correspond in turn to $k = 2, \ldots, 8$. It can be seen from the figure that the relative reduction increases symmetrically as p goes from 0.5 to 0 and 1. This indicates that the advantage of RSS is greater for estimating the densities at the tails of the distribution.

To end this section, we present some results of a simulation study. The following three distributions are used in the simulation: the standard normal distribution $N(0,1)$, the gamma distribution with shape parameter $\beta = 3$ and the extreme value distribution with location parameter $\mu = 0$ and scale parameter $\sigma = 1$. The following combinations of set size k and cycle number m are considered: (1) $k = 6, m = 4$, (2) $k = 6, m = 8$, (3) $k = 8, m = 3$ and

Table 2.4. The comparison of the MISE between the RSS and SRS kernel estimates

Distribution	k	m	MISE.SRS	MISE.RSS	Rel. Eff.
Normal	6	4	0.0182(0.016)	0.0112(0.009)	1.625
(0, 1)	6	8	0.0102(0.008)	0.0067(0.005)	1.522
	8	3	0.0182(0.016)	0.0100(0.008)	1.820
	8	6	0.0102(0.008)	0.0062(0.004)	1.645
Gamma	6	4	0.0123(0.010)	0.0086(0.006)	1.430
$\beta = 3$	6	8	0.0074(0.006)	0.0053(0.003)	1.396
	8	3	0.0123(0.010)	0.0078(0.005)	1.577
	8	6	0.0074(0.006)	0.0049(0.003)	1.510
Extreme Value	6	4	0.0171(0.015)	0.0113(0.008)	1.513
$\mu = 0$	6	8	0.0098(0.007)	0.0068(0.004)	1.441
$\sigma = 1$	8	3	0.0171(0.015)	0.0105(0.007)	1.629
	8	6	0.0098(0.007)	0.0065(0.004)	1.508

(4) $k = 8, m = 6$. For each combination of k and m and each distribution, 5000 SRS samples of size mk and 5000 RSS samples with set size k and cycle number m are generated using the random number generating functions in Splus 4.5. For each of these samples, a kernel estimate of the underlying density function is obtained. The Epanechnikov kernel, which is optimal in the sense that it minimizes $C(K)$ among certain class of kernels, is used in all the estimates. The Epanechnikov kernel is given by

$$K(x) = \begin{cases} \frac{3}{4\sqrt{5}}(1 - \frac{x^2}{5}), & -\sqrt{5} \le x \le \sqrt{5}, \\ 0, & \text{otherwise.} \end{cases}$$

The bandwidth h is determined by

$$h = \frac{5}{4} C(K) \left[\frac{3}{8\sqrt{\pi}} \right]^{-1/5} A n^{-1/5},$$

where A = min{standard deviation of the sample, interquartile range of the sample /1.34}. Since A differs from sample to sample, the bandwidths h used in the estimates are not exactly the same. However, they have the same order $O(n^{-1/5})$. For each estimate the integrated square error (ISE) $\int [\hat{f}(x) - f(x)]^2 dx$ is computed by numerical method. Then the average and the standard deviation of the ISEs of the 5000 RSS estimates are computed. The same is done to the 5000 SRS estimates. The averages of the ISEs are taken as the estimate of the MISEs and are compared between the RSS and the SRS estimates. The estimated MISEs are given in Table 2.4. The numbers in the parentheses are the standard deviations. The improvement on the MISE by using RSS is quite significant. The assertions we made from (2.24) manifest in the table. For fixed k, the relative efficiency decreases slowly as the sample size increases. On the other hand, for fixed sample size, the increment in relative efficiency is much faster when k increases.

The other methods of density estimation can also be investigated in a similar manner as we have done in this section. Because of the fact that a ranked-set sample indeed contains more information than a simple random sample of the same size, we expect that for all such methods the RSS version will provide a better estimate than the SRS version.

2.8 M-estimates with RSS data

The idea of M-estimates arises out of concern on the robustness of statistical procedures. For example, the usual estimate of the population mean, namely the sample mean, is not robust if the underlying distribution is heavily tailed. This problem does not go away in RSS. In this section, we investigate the properties of the RSS M-estimates including their asymptotic distribution and their efficiency relative to their counterparts with SRS. The RSS estimates are defined and their asymptotic properties are dealt with in Section 2.8.1. The relative efficiency of the RSS M-estimates is discussed in Section 2.8.2.

2.8.1 The RSS M-estimates and their asymptotic properties

Let $\psi(x)$ be an appropriate function. Define the functional $T(F)$ over all distribution functions as the solution of $\lambda_F(t) = \int \psi(x-t)dF(x) = 0$, if exists. In generic notation, if F is an unknown distribution function and $\hat{F}$ is an appropriate estimate of F, then $T(\hat{F})$ is called an M-estimate of $T(F)$. Let $\hat{T}_n = T(\hat{F}_{\text{RSS}})$, i.e., $\lambda_{\hat{F}_{\text{RSS}}}(\hat{T}_n) = 0$, which defines the RSS M-estimate of $T(F)$.

The following lemma is used later.

Lemma 2.15. *Suppose that $\psi(x)$ is an odd function and F is a symmetric location distribution, then the population mean μ is a solution of $\lambda_F(t) = 0$, i.e., $\mu = T(F)$, and, further, μ satisfies*

$$\int_{-\infty}^{+\infty} \psi(x-t)dF_{(r)}(x) + \int_{-\infty}^{+\infty} \psi(x-t)dF_{(k-r+1)}(x) = 0.$$

The following theorem gives conditions under which the RSS M-estimate exists and is also consistent.

Theorem 2.16. *Suppose that $\psi(x)$ is odd, continuous and either monotone or bounded, and that F is a symmetric location distribution. Then there is a solution sequence $\{\hat{T}_n\}$ of $\lambda_{\hat{F}_{RSS}}(t) = 0$ such that $\{\hat{T}_n\}$ converges to μ with probability 1.*

There are other conditions on ψ so that Theorem 2.16 holds. However, since the ψ's in practical applications satisfy the conditions in Theorem 2.16, we will concentrate on the ψ's satisfying these conditions. In particular, we will later consider the following two ψ functions. The first ψ is given by

$$\psi_1 = \begin{cases} -1.5, x < -1.5, \\ x, \quad |x| \leq 1.5, \\ 1.5, \quad x > 1.5. \end{cases}$$

The corresponding M-estimator is a type of Winsorized mean. The other ψ is a smoothed "Hampel" given by

$$\psi_2 = \begin{cases} \sin(x/2.1), & |x| < 2.1\pi, \\ 0, & |x| \geq 2.1\pi. \end{cases}$$

Let

$$A_{(r)}(t) = \int \psi^2(x-t)dF_{(r)}(x) - [\int \psi(x-t)dF_{(r)}(x)]^2,$$

$$\lambda_F'(T(F)) = \left. \frac{d \int \psi(x-t)dF(x)}{dt} \right|_{t=T(F)}.$$

In the following, we give three sets of conditions each of which, together with the conditions on F given in Theorem 2.16, guarantees the asymptotic normality of the sequence $\hat{T}_n$.

A1 $\psi(x)$ is odd and monotone; $\lambda_F(t)$ is differentiable at $t = \mu$, with $\lambda_F'(\mu) \neq 0$; $\int \psi^2(x-t)dF(x)$ is finite for t in a neighborhood of μ and is continuous at $t = \mu$.

A2 $\psi(x)$ is odd, continuous and satisfies $\lim_{t\to\mu} \parallel \psi(\cdot,t) - \psi(\cdot,\mu) \parallel_V = 0$; $\int \psi^2(x-t)dF(x) < \infty$ and $\lambda_F(t)$ is differentiable at $t = \mu$, with $\lambda_F'(\mu) \neq 0$.

A3 $\psi(x)$ is odd and uniformly continuous; $\int \partial\psi(x-t)/\partial t|_{t=\mu}dF(x)$ is finite and nonzero; $\int \psi^2(x-\mu)dF(x) < \infty$.

Theorem 2.17. *Assume that F is a symmetric location distribution. Then, under either* (**A1**), (**A2**) *or* (**A3**),

$$\sqrt{n}(\hat{T}_n - \mu) \to N(0, \sigma^2_{RSS}(F)), \tag{2.25}$$

in distribution, where under (**A1**) *and* (**A2**),

$$\sigma^2_{RSS}(F) = \frac{1}{k}\sum_{r=1}^{k} A_{(r)}(\mu) \Big/ \left[\frac{1}{k}\sum_{r=1}^{k} \lambda'_{F_{(r)}}(\mu)\right]^2,$$

and under (**A3**),

$$\sigma^2_{RSS}(F) = \frac{1}{k}\sum_{r=1}^{k} A_{(r)}(\mu) \Big/ \left[\frac{1}{k}\sum_{r=1}^{k} \int \partial\psi(x-t)/\partial t|_{t=\mu}dF_{(r)}(x)\right]^2.$$

The results in this section are straightforward extensions of the corresponding results in SRS. A sketch of the proof can be found in Zhao and Chen [175].

2.8.2 The relative efficiency of the RSS M-estimates

In this section, we deal with the ARE of RSS M-estimates with respect to SRS M-estimates. The SRS M-estimate of μ is given by $\tilde{T}_n = T(F_n)$ where F_n is the empirical distribution of a simple random sample of size n. The SRS M-estimate has asymptotically a normal distribution with mean μ and variance $\sigma^2_{\text{SRS}}(F)$ given, depending on the assumptions on $\psi(x)$, by either

$$\int \psi^2(x-\mu)dF(x)/[\lambda'_F(\mu)]^2$$

or

$$\int \psi^2(x-\mu)dF(x)/[\int \partial\psi(x-t)/\partial t|_{t=\mu}dF(x)]^2.$$

Note that, because of the fundamental equality,

$$\frac{1}{k}\sum_{r=1}^{k}\lambda'_{F_{(r)}}(\mu) = \lambda'_F(\mu);$$

$$\frac{1}{k}\sum_{r=1}^{k}\int \partial\psi(x-t)/\partial t|_{t=\mu}dF_{(r)}(x) = \int \partial\psi(x-t)/\partial t|_{t=\mu}dF(x);$$

$$\frac{1}{k}\sum_{r=1}^{k}\int \psi^2(x-t)dF_{(r)}(x) = \int \psi^2(x-t)dF(x).$$

Hence the ARE of the RSS M-estimate is given by

$$ARE(\hat{T}_n, \tilde{T}_n) = \frac{\int \psi^2(x-\mu)dF(x)}{\int \psi^2(x-\mu)dF(x) - \frac{1}{k}\sum_{r=1}^{k}[\int \psi(x-\mu)dF_{(r)}(x)]^2}.$$

It is obvious that the ARE is greater than 1. The ARE's of the RSS M-estimates with ψ_1 and ψ_2 for the $Cauchy(0,1)$, $N(0,1)$ and some contaminated normal distributions are given, respectively, in Table 2.5 and Table 2.6. It is evident that, as expected, RSS is much more efficient than SRS.

2.9 Appendix: Technical details for Section 2.4*

We first state and prove a general result. Let $Y_1, \cdots, Y_n$ be independent random variables with $E(Y_i) = \nu_i$, $\text{Var}(Y_i) = \eta_i^2$, $\gamma_i = E(Y_i - \nu_i)^3$, and $\kappa_i = E(Y_i - \nu_i)^4 - 3\eta_i^4$, $i = 1, \cdots, n$. Denote $\boldsymbol{Y} = (Y_1, \cdots, Y_n)'$, $\Delta = diag(\eta_1^2, \cdots, \eta_n^2)$, $\boldsymbol{\nu} = (\nu_1, \cdots, \nu_n)'$. Let Q be an $n \times n$ symmetric matrix.

Table 2.5. The ARE's of RSS w.r.t. SRS with ψ_1 for Cauchy, Normal and some contaminated Normal distributions

Dist.	Cauchy(0, 1)	N(0, 1)	0.9N(0, 1) +0.1N(0, 9)	0.7N(0, 1) +0.3N(0, 9)	0.5N(0, 1) +0.5N(0, 9)
k=2	1.4868	1.4949	1.4958	1.4939	1.4858
k=3	1.9651	1.9866	1.9888	1.9838	1.9625
k=4	2.4364	2.4757	2.4792	2.4700	2.4325
k=5	2.9016	2.9627	2.9673	2.9527	2.8975
k=6	3.3613	3.4481	3.4533	3.4321	3.3582
k=7	3.8162	3.9320	3.9372	3.9085	3.8154
k=8	4.2668	4.4145	4.4193	4.3823	4.2696
k=9	4.7136	4.8959	4.8998	4.8538	4.7212
k=10	5.1570	5.3762	5.3789	5.3232	5.1705

Table 2.6. The ARE's of RSS w.r.t. SRS with ψ_2 for Cauchy, Normal and some contaminated Normal distributions

Dist.	Cauchy(0, 1)	N(0, 1)	0.9N(0, 1) +0.1N(0, 9)	0.7N(0, 1) +0.3N(0, 9)	0.5N(0, 1) +0.5N(0, 9)
k=2	1.3038	1.4878	1.4767	1.4584	1.4380
k=3	1.5374	1.9676	1.9389	1.8922	1.8412
k=4	1.7327	2.4422	2.3886	2.3031	2.2157
k=5	1.9033	2.9131	2.8271	2.6930	2.5658
k=6	2.0570	3.3812	3.2554	3.0639	2.8952
k=7	2.1989	3.8471	3.6740	3.4176	3.2074
k=8	2.3321	4.3113	4.0836	3.7559	3.5051
k=9	2.4589	4.7741	4.4845	4.0805	3.7907
k=10	2.5807	5.2358	4.8772	4.3928	4.0662

Lemma 2.18.

$$E(\boldsymbol{Y}'Q\boldsymbol{Y}) = tr[Q(\Delta + \boldsymbol{\nu}\boldsymbol{\nu}'] \tag{2.26}$$

$$\begin{aligned} Var((\boldsymbol{Y}'Q\boldsymbol{Y}) &= \sum_{i=1}^{n} Q_{ii}^2\kappa_i + 2tr[(Q\Delta)^2] + 4\boldsymbol{\nu}'Q\Delta Q\boldsymbol{\nu} \\ &\quad +4\sum_{i=1}^{n} Q_{ii}\gamma_i(\sum_{l=1}^{n} Q_{il}\nu_l). \end{aligned} \tag{2.27}$$

Proof. (2.26) is obvious. To prove (2.27), write

$$\boldsymbol{Y}'Q\boldsymbol{Y} = (\boldsymbol{Y} - \boldsymbol{\nu})'Q(\boldsymbol{Y} - \boldsymbol{\nu}) + 2\boldsymbol{\nu}'Q(\boldsymbol{Y} - \boldsymbol{\nu}) + \boldsymbol{\nu}'Q\boldsymbol{\nu}.$$

Then

$$\mathrm{Var}(\boldsymbol{Y}'Q\boldsymbol{Y}) = \mathrm{Var}[(\boldsymbol{Y} - \boldsymbol{\nu})'Q(\boldsymbol{Y} - \boldsymbol{\nu})] + 4\mathrm{Var}[\boldsymbol{\nu}'Q(\boldsymbol{Y} - \boldsymbol{\nu})]$$

$$+4\mathrm{Cov}[(\boldsymbol{Y}-\boldsymbol{\nu})'Q(\boldsymbol{Y}-\boldsymbol{\nu}), \boldsymbol{\nu}'Q(\boldsymbol{Y}-\boldsymbol{\nu})].$$

The first term can be simplified as

$$\begin{aligned}
&\mathrm{Var}[(\boldsymbol{Y}-\boldsymbol{\nu})'Q(\boldsymbol{Y}-\boldsymbol{\nu})] \\
&= \sum_{i=1}^{n}\sum_{j=1}^{n}\sum_{k=1}^{n}\sum_{l=1}^{n} Q_{ij}Q_{kl}\mathrm{Cov}[(Y_i-\nu_i)(Y_j-\nu_j), (Y_k-\nu_k)(Y_l-\nu_l)] \\
&= \sum_{i=1}^{n} Q_{ii}^2 \mathrm{Var}[(Y_i-\nu_i)^2] + 2\sum_{i\neq j=1}^{n} Q_{ij}^2 \mathrm{Var}[(Y_i-\nu_i)(Y_j-\nu_j)] \\
&= \sum_{i=1}^{n} Q_{ii}^2(\kappa_i + 2\eta_i^4) + 2\sum_{i\neq j=1}^{n} Q_{ij}^2\eta_i^2\eta_j^2 \\
&= \sum_{i=1}^{n} Q_{ii}^2\kappa_i + 2\sum_{i=1}^{n}\sum_{j=1}^{n} Q_{ij}^2\eta_i^2\eta_j^2 \\
&= \sum_{i=1}^{n} Q_{ii}^2\kappa_i + 2tr[(Q\Delta)^2].
\end{aligned}$$

The second term simplifies as

$$4\mathrm{Var}[\boldsymbol{\nu}'Q(\boldsymbol{Y}-\boldsymbol{\nu})] = 4\boldsymbol{\nu}'Q\Delta Q\boldsymbol{\nu}.$$

Finally, the third term can be simplified as

$$\begin{aligned}
&4\mathrm{Cov}[(\boldsymbol{Y}-\boldsymbol{\nu})'Q(\boldsymbol{Y}-\boldsymbol{\nu}), \boldsymbol{\nu}'Q(\boldsymbol{Y}-\boldsymbol{\nu})] \\
&= 4\sum_{i=1}^{n}\sum_{j=1}^{n}\sum_{k=1}^{n}\sum_{l=1}^{n} Q_{ij}Q_{kl}\mathrm{Cov}[(Y_i-\nu_i)(Y_j-\nu_j), (Y_k-\nu_k)\nu_l] \\
&= 4\sum_{i=1}^{n}\sum_{l=1}^{n} Q_{ii}Q_{il}\gamma_i\nu_l \\
&= 4\sum_{i=1}^{n} Q_{ii}\gamma_i[\sum_{l=1}^{n} Q_{il}\nu_l].
\end{aligned}$$

The lemma then follows.

We now apply the lemma to the special case with Q given by the form

$$Q = A \otimes [I_m - \frac{\mathbf{1}_m\mathbf{1}_m'}{m}] + D \otimes \frac{\mathbf{1}_m\mathbf{1}_m'}{m},$$

where A and D are arbitrary symmetric matrices. For convenience, let $K_1 = \mathbf{1}_m\mathbf{1}_m'/m$ and $K_0 = I_m - K_1$, and, similarly, let $J_1 = \mathbf{1}_k\mathbf{1}_k'/k$ and $J_0 = I_k - J_1$. Let

$$\begin{aligned}
&\boldsymbol{\mu} = (\mu_{[1]}, \mu_{[2]}, \ldots, \mu_{[k]})', \ \Sigma = \mathrm{diag}(\sigma_{[1]}^2, \sigma_{[2]}^2, \ldots, \sigma_{[k]}^2) \\
&\boldsymbol{\nu} = \boldsymbol{\mu} \otimes \mathbf{1}_m, \ \Delta = \Sigma \otimes I_m.
\end{aligned}$$

Also, we denote by $[D\boldsymbol{\mu}]_r$ the rth element of the vector $D\boldsymbol{\mu}$. Then we have

Lemma 2.19.

$$E(\boldsymbol{X}'Q\boldsymbol{X}) = tr[((m-1)A + D)\Sigma + mD\boldsymbol{\mu}\boldsymbol{\mu}'], \tag{2.28}$$

$$\begin{aligned} Var(\boldsymbol{X}'Q\boldsymbol{X}) = & \frac{1}{m}\sum_{r=1}^{k}[(m-1)a_{rr} + d_{rr}]^2\kappa^*_{[r]} \\ & +2[(m-1)tr(A\Sigma)^2 + tr(D\Sigma)^2] + 4m(D\boldsymbol{\mu})'\Sigma D\boldsymbol{\mu} \\ & +\frac{4}{m}\sum_{r=1}^{k}[(m-1)a_{rr} + d_{rr}]\gamma^*_{[r]}[D\boldsymbol{\mu}]_r. \end{aligned} \tag{2.29}$$

Proof. Using the special structure of $\boldsymbol{\nu}$, Q and Δ, we get

$$\begin{aligned} & \mathrm{tr}[Q(\Delta + \boldsymbol{\nu}\boldsymbol{\nu}')] \\ & = \mathrm{tr}[(A \otimes K_0 + D \otimes K_1)(\Sigma \otimes \boldsymbol{I}_m + \boldsymbol{\mu} \otimes \mathbf{1}_m(\boldsymbol{\mu} \otimes \mathbf{1}_m)')] \\ & = \mathrm{tr}[A\Sigma \otimes K_0 + D\Sigma \otimes K_1] + \boldsymbol{\mu}'A\boldsymbol{\mu}\mathbf{1}_m'K_0\mathbf{1}_m + \boldsymbol{\mu}'D\boldsymbol{\mu}\mathbf{1}_m'K_1\mathbf{1}_m \\ & = (m-1)\mathrm{tr}(A\Sigma) + \mathrm{tr}(D\Sigma) + m\boldsymbol{\mu}'D\boldsymbol{\mu}. \end{aligned}$$

since $\mathrm{tr}(A \otimes D) = \mathrm{tr}(A)\mathrm{tr}(D)$, $\mathbf{1}_m'K_0\mathbf{1}_m = 0$ and $\mathbf{1}_m'K_1\mathbf{1}_m = m$. Hence (2.28) is proved.

Note that, in view of the special structure of Q and $\boldsymbol{X}$,

$$\begin{aligned} & \kappa_j = \kappa^*_{[r]},\ \gamma_j = \gamma^*_{[r]},\ Q_{jj} = \frac{(m-1)a_{rr}}{m} + \frac{d_{rr}}{m}, \\ & j = (r-1)m+1, \cdots, rm, \quad r = 1, \cdots, k. \end{aligned} \tag{2.30}$$

Hence, we readily obtain

$$\sum_{j=1}^{n} Q_{jj}^2\kappa_j = m\sum_{r=1}^{k}[\frac{(n-1)a_{rr}}{m} + \frac{d_{rr}}{m}]^2\kappa^*_{[r]}, \tag{2.31}$$

where $n = mk$. Next, note that in view of idempotence of K_0 and K_1 and the fact that $K_0K_1 = \mathbf{0}$, we get

$$\begin{aligned} \mathrm{tr}(Q\Delta)^2 & = \mathrm{tr}[(A \otimes K_0 + D \otimes K_1)(\Sigma \otimes \boldsymbol{I}_m)]^2 \\ & = \mathrm{tr}[(A\Sigma \otimes K_0 + D\Sigma \otimes K_1)^2] \\ & = \mathrm{tr}[A\Sigma A\Sigma \otimes K_0 + D\Sigma D\Sigma \otimes K_1] \\ & = (m-1)\mathrm{tr}(A\Sigma)^2 + \mathrm{tr}(D\Sigma)^2. \end{aligned}$$

Moreover, since

$$Q\boldsymbol{\nu} = (A \otimes K_0 + D \otimes K_1)(\boldsymbol{\mu} \otimes \mathbf{1}_m) = D\boldsymbol{\mu} \otimes \mathbf{1}_m,$$

using the special structure of Δ, we get

$$(Q\boldsymbol{\nu})'\Delta(Q\boldsymbol{\nu}) = m(D\boldsymbol{\mu})'\Sigma(D\boldsymbol{\mu}).$$

Finally, using (2.30),

$$\sum_{j=1}^{n} Q_{jj}\gamma_j[\sum_{l=1}^{n} Q_{jl}\nu_l] = m\sum_{r=1}^{k}[\frac{(m-1)a_{rr}}{m} + \frac{d_{rr}}{m}]\gamma^*_{[r]}[D\boldsymbol{\mu}]_r.$$

Hence (2.29) is proved.

We are now in a position to prove Theorem 2.4.

Proof. (a) By (2.28),

$$E[\boldsymbol{X}'Q\boldsymbol{X}] = \mathrm{tr}[[(m-1)A + D]\Sigma + mD\boldsymbol{\mu}\boldsymbol{\mu}'].$$

On the other hand,

$$\sigma^2 = \frac{1}{k}\mathrm{tr}[\Sigma + J_0\boldsymbol{\mu}\boldsymbol{\mu}'].$$

Therefore, $\hat{\sigma}^2$ is unbiased for σ^2 if and only if $\mathrm{tr}[[(m-1)A+D-(1/k)\boldsymbol{I}]\Sigma] = 0$ for all diagonal Σ and $\boldsymbol{\mu}'[mD-(1/k)J_0]\boldsymbol{\mu} = 0$ for all possible values of $\boldsymbol{\mu}$. Hence the proof of (a).

(b) Using the condition of unbiasedness, it is clear that the only term in $\mathrm{Var}(\boldsymbol{X}'Q\boldsymbol{X})$ which depends on the off-diagonal elements of the matrix A is given by $\mathrm{tr}(A\Sigma)^2 = \sum_{r=1}^{k}\sum_{s=1}^{k} a_{rs}^2\sigma_{[r]}^2\sigma_{[s]}^2$. Obviously, a_{rr}'s are fixed from the unbiasedness condition, and the unique choice of a_{rs}'s for $r \neq s$ which makes $\mathrm{Var}(\boldsymbol{X}'Q\boldsymbol{X})$ a minimum is given by $a_{rs} = 0$ for all $r \neq s$. Hence the proof.

2.10 Bibliographic notes

The general framework of consistent ranked set sampling and an unified treatment on the estimation of means and smooth-function-of-means were given in Bai and Chen [6]. The minimum variance non-negative unbiased estimate of variance was developed by Perron and Sinha [132]. A similar estimate of variance was considered by MacEachern et al. [94]. The asymptotic pivot based on RSS for the tests and confidence intervals of a population mean was dealt with by Chen and Sinha [43]. The estimation of quantiles was treated in Chen [35]. Density estimation was considered in Chen [34]. The M-estimates were studied in full detail by Zhao and Chen [175]. The early research on ranked set sampling was concentrated on the non-parametric setting. Besides the seminal paper of McIntyre [96], earlier works include Halls and Dell [58], Takahasi and Wakimoto [167], Takahasi [163], [164], Dell and Clutter [50], Stokes [156],[157], [159], [158] etc.. Other aspects of ranked set sampling in the non-parametric setting were explored in the literature as well. Stokes and Sager [162] gave a characterization of ranked set sample and considered the estimation of distribution function. Kvam and Samaniego [83] considered the inadmissibility of empirical averages as estimators in ranked set sampling. Kvam and Samaniego

[85] and Huang [63] considered the nonparametric maximum likelihood estimation based on ranked set samples. Samawi and Muttlak [143] considered estimation of ratio using ranked set sample. Patil et al. [124] [129] dealt with ranked set sampling for finite populations. Presnell and Bohn [134] tackled U-statistics in ranked set sampling.

3

Balanced Ranked Set Sampling II: Parametric

We turn to balanced parametric RSS in this chapter. It is assumed that the underlying distribution is known to belong to a certain distribution family up to some unknown parameters. From an information point of view, intuitively, the amount of information contained in a ranked set sample should be larger than that contained in a simple random sample of the same size, since a ranked set sample contains not only the information carried by the measurements but also the information carried by the ranks. We shall deal with the Fisher information of an RSS sample and make this assertion rigorous. In Section 3.1, we consider the Fisher information of a ranked set sample first for the special case of perfect ranking and then for general cases when the assumption of perfect ranking is dropped. In the case of perfect ranking, we derive the result that the information matrix based on a balanced ranked set sample is the sum of the information matrix based on a simple random sample of the same size and a positive definite information gain matrix. In general cases, it is established that the information matrix based on a balanced ranked set sample minus the information matrix based on a simple random sample of the same size is always non-negative definite. It is also established that the positive-definiteness of the difference holds as long as ranking in the RSS is not a purely random permutation. Conditions for the difference of the two information matrices to be of full rank are also given. In Section 3.2, we discuss maximum likelihood estimation (MLE) based on ranked set samples and its relative efficiency with respect to MLE based on simple random samples. The MLE based on RSS, compared with other estimates based on RSS, has the same optimality properties as those of the MLE based on SRS such as the asymptotic normality and the attainment of Cramér-Rao lower bound as long as certain regularity conditions are satisfied. The relative efficiency of the MLE based on RSS is greater than 1 in general. Specific values of the relative efficiency for the estimation of mean and variance for Normal, Gamma and Weibull families are given. The relative efficiency for the estimation of mean is much greater than that for the estimation of variance. In Section 3.3, we consider best linear unbiased estimation in the context of RSS, which provides an alternative to MLE for location-scale families. The best linear unbiased estimation is not as efficient as the MLE in general, but is much easier to compute. The regularity conditions and the existence of the Fisher information of RSS samples are discussed in Section 3.4. Section s 3.1.2 and 3.4 are technically more involved. The readers who have difficulty with these sections can skip them without affecting their understanding of the materials in later chapters.

3.1 The Fisher information of ranked set samples

Let $f(x;\boldsymbol{\theta})$ and $F(x;\boldsymbol{\theta})$ denote, respectively, the density function and cumulative distribution function of the population under study, where $\boldsymbol{\theta} = (\theta_1, \cdots, \theta_q)^T$ is a vector of unknown parameters taking values in an open set Θ of R^q. Let $f_{(r)}(x;\boldsymbol{\theta})$ and $F_{(r)}(x;\boldsymbol{\theta})$ denote, respectively, the density function and the cumulative distribution function of the rth order statistic of a simple random sample of size k from F. Let a balanced ranked set sample with set size k and cycle number m be denoted by $\boldsymbol{X}_n = \{X_{(r)i}, r = 1, \cdots, k;\ i = 1, \cdots, m\}$, where $n = mk$. If ranking is imperfect, we replace, in the notations above, the parentheses by brackets, i.e., $f_{(r)}, F_{(r)}$ and $X_{(r)}$ are replaced by $f_{[r]}, F_{[r]}$ and $X_{[r]}$, respectively. Denote by $I(\boldsymbol{\theta})$ the Fisher information matrix of a single random observation from F and by $I_{\text{RSS}}(\boldsymbol{\theta}, n)$ the Fisher information matrix of a ranked set sample with size $n = mk$. In the case of perfect ranking, we shall establish a relationship between $I_{\text{RSS}}(\boldsymbol{\theta}, n)$ and $I(\boldsymbol{\theta})$. In general, we shall make a comparison between $I_{\text{RSS}}(\boldsymbol{\theta}, n)$ and $nI(\boldsymbol{\theta})$, the latter being the Fisher information based on an SRS sample of size n. The major result of this section is that $I_{\text{RSS}}(\boldsymbol{\theta}, n) - nI(\boldsymbol{\theta})$ is always non-negative definite whether ranking is perfect or not. For matrices A and B, we use the notation $A \geq 0$ and $A \geq B$ to indicate that A and $A - B$ are non-negative definite. For positive definiteness, "$\geq$" is replaced by "$>$".

3.1.1 The Fisher information of ranked set samples when ranking is perfect

Under the assumption of perfect ranking, the likelihood of $\boldsymbol{\theta}$ based on the ranked set sample $\boldsymbol{X}_n$ is given by

$$L(\boldsymbol{X}_n, \boldsymbol{\theta}) = \prod_{r=1}^{k} \prod_{i=1}^{m} f_{(r)}(X_{(r)i}; \boldsymbol{\theta})$$

where

$$f_{(r)}(x;\boldsymbol{\theta}) = \frac{k!}{(r-1)!(k-r)!} F^{r-1}(x;\boldsymbol{\theta})[1 - F(x;\boldsymbol{\theta})]^{k-r} f(x;\boldsymbol{\theta}). \qquad (3.1)$$

We have the following theorem.

Theorem 3.1. *Under certain regularity conditions (see the Appendix) and the assumption of perfect ranking, we have*

$$I_{RSS}(\boldsymbol{\theta}) = mkI(\boldsymbol{\theta}) + mk(k-1)\Delta(\boldsymbol{\theta}), \tag{3.2}$$

where

$$\Delta(\boldsymbol{\theta}) = E\left\{\frac{\boldsymbol{D}F(X;\boldsymbol{\theta})[\boldsymbol{D}F(X;\boldsymbol{\theta})]^T}{F(X;\boldsymbol{\theta})(1-F(X;\boldsymbol{\theta}))}\right\},$$

with $\boldsymbol{D} = (D_1, \cdots, D_q) = (\frac{\partial}{\partial\theta_1}, \cdots, \frac{\partial}{\partial\theta_q})^T$. *The matrix* $\Delta(\boldsymbol{\theta})$ *is obviously non-negative definite. Furthermore, it is positive definite unless there is a nonzero vector* $\boldsymbol{b} = \boldsymbol{b}(\boldsymbol{\theta}) = (b_1, \cdots, b_q)^T$ *such that for almost all* $x[f(x;\boldsymbol{\theta})]$,

$$\boldsymbol{b}'\boldsymbol{D}f(x;\boldsymbol{\theta}) = 0. \tag{3.3}$$

Remark: If $I(\boldsymbol{\theta}) > 0$, then it is impossible for (3.3) to hold since then (3.3) implies that $\boldsymbol{b}'I(\boldsymbol{\theta})\boldsymbol{b} = 0$. Therefore, by Theorem 3.1, the information matrix of the ranked set sample is strictly larger than that of the simple random sample if $I(\boldsymbol{\theta}) > 0$.

Since the matrix $\Delta(\boldsymbol{\theta})$ is always non-negative definite, it is referred to as the information gain matrix of the RSS. The following lemma is used in the proof of Theorem 3.1. The lemma has its own interest as well.

Lemma 3.2. *Let* $X_{(k:r)}$ *be the r-th order statistic of a random sample of size k from a distribution with continuous distribution function* $F(x)$. *Then, for any function* $G(x)$, *we have*

$$E\sum_{r=1}^{k}(r-1)\frac{G(X_{(k:r)})}{F(X_{(k:r)})} = k(k-1)EG(X), \tag{3.4}$$

$$E\sum_{r=1}^{k}(k-r)\frac{G(X_{(k:r)})}{1-F(X_{(k:r)})} = k(k-1)EG(X),$$

provided that $EG(X)$ *exists, where* X *is a random variable with distribution function* $F(x)$.

PROOF: We have

$$\begin{aligned}
&E\sum_{r=1}^{k}(r-1)\frac{G(X_{(k:r)})}{F(X_{(k:r)})} \\
&= \sum_{r=1}^{k}(r-1)\int\frac{k!}{(r-1)!(k-r)!}\frac{G(x)}{F(x)}F^{r-1}(x)(1-F(x))^{k-r}dF(x) \\
&= k(k-1)\int G(x)dF(x) = k(k-1)EG(X).
\end{aligned}$$

The second equality is a dual version of (3.4) and its proof is omitted.

Now we give a proof of Theorem 3.1.
Proof of Theorem 3.1: It suffices to prove in the case $m = 1$. For the sake of convenience, we drop the second subscript in $X_{(r)1}$ and write it as $X_{(r)}$. Let

$$\ell_1(\boldsymbol{\theta}) = \sum_{r=1}^{k}[(r-1)\log(F(X_{(r)};\boldsymbol{\theta})) + (k-r)\log(1-F(X_{(r)};\boldsymbol{\theta})],$$

$$\ell_2(\boldsymbol{\theta}) = \sum_{r=1}^{k}\log(f(X_{(r)};\boldsymbol{\theta}).$$

It follows from (3.1) that $\log L(\boldsymbol{X}_N;\boldsymbol{\theta}) = \ell_1(\boldsymbol{\theta}) + \ell_2(\boldsymbol{\theta})$, and hence that

$$I_{\text{RSS}}(\boldsymbol{\theta}, N) = -E\left(\nabla\ell_1(\boldsymbol{\theta})\right) - E\left(\nabla\ell_2(\boldsymbol{\theta})\right),$$

where $\nabla = \boldsymbol{D}\boldsymbol{D}^T = \left[\frac{\partial^2}{\partial\theta_i\partial\theta_j}, i,j = 1,\cdots,q\right]$. It is easy to see from the fundamental equality that

$$-E\left(\nabla\ell_2(\boldsymbol{\theta})\right) = kI(\boldsymbol{\theta}).$$

We also have

$$\begin{aligned}\nabla_{ij}\ell_1(\boldsymbol{\theta}) &= \sum_{r=1}^{k}(r-1)\left[\frac{F''_{ij}(X_{(r)};\boldsymbol{\theta})}{F(X_{(r)};\boldsymbol{\theta})} - \frac{F'_i(X_{(r)};\boldsymbol{\theta})F'_j(X_{(r)};\boldsymbol{\theta})}{F^2(X_{(r)};\boldsymbol{\theta})}\right] \\ &\quad - \sum_{r=1}^{k}(k-r)\left[\frac{F''_{ij}(X_{(r)};\boldsymbol{\theta})}{1-F(X_{(r)};\boldsymbol{\theta})} + \frac{F'_i(X_{(r)};\boldsymbol{\theta})F'_j(X_{(r)};\boldsymbol{\theta})}{(1-F(X_{(r)};\boldsymbol{\theta}))^2}\right],\end{aligned}$$

where $F''_{ij} = \nabla_{ij}F$ and $F'_i = D_iF$. By Lemma 3.2, the (i,j)-th element of $\Delta(\boldsymbol{\theta})$ is equal to

$$\begin{aligned}\Delta(\boldsymbol{\theta})_{ij} &= -E\left(\nabla_{ij}\ell_1(\boldsymbol{\theta})\right) \\ &= k(k-1)\left[-EF''_{ij}(X;\boldsymbol{\theta}) + E\frac{F'_i(X;\boldsymbol{\theta})F'_j(X;\boldsymbol{\theta})}{F(X;\boldsymbol{\theta})}\right. \\ &\quad \left.+EF''_{ij}(X;\boldsymbol{\theta}) + E\frac{F'_i(X;\boldsymbol{\theta})F'_j(X;\boldsymbol{\theta})}{1-F(X;\boldsymbol{\theta})}\right] \\ &= E\frac{F'_i(X;\boldsymbol{\theta})F'_j(X;\boldsymbol{\theta})}{F(X;\boldsymbol{\theta})(1-F(X;\boldsymbol{\theta}))}.\end{aligned}$$

As to the last conclusion of the theorem, if $\Delta(\boldsymbol{\theta})$ is not positive definite, then there is a nonzero vector $\boldsymbol{b} = (b_1,\cdots,b_q)^T$ such that $\boldsymbol{b}^T\Delta(\boldsymbol{\theta})\boldsymbol{b} = 0$. Notice that

$$\boldsymbol{b}^T\Delta(\boldsymbol{\theta})\boldsymbol{b} = \int\frac{(\boldsymbol{b}'DF(x;\boldsymbol{\theta}))^2}{F(x;\boldsymbol{\theta})(1-F(x;\boldsymbol{\theta}))}f(x;\boldsymbol{\theta})dx.$$

We conclude that $\boldsymbol{b}^TDF(x;\boldsymbol{\theta}) = 0$ for almost all $x[f(x;\boldsymbol{\theta})]$. We further have

$$b^T DF(x;\theta) = \int_{-\infty}^{x} b^T Df(y;\theta)dy.$$

We then obtain that $b^T Df(x;\theta) = 0$ for almost all $x[f(x;\theta)]$. The proof of the theorem is then complete.

Next, we consider two important special cases of Theorem 3.1.

Location-scale family. Let $F(x;\theta) = F(\frac{x-\mu}{\lambda})$, where $\theta = (\mu, \lambda)^T$. Here, the functional form of the distribution function F is assumed to be known. For this distribution family, we have

$$\frac{\partial F\left(\frac{X-\mu}{\lambda}\right)}{\partial \mu} = -\frac{1}{\lambda} f(\frac{X-\mu}{\lambda})$$
$$\frac{\partial F\left(\frac{X-\mu}{\lambda}\right)}{\partial \lambda} = -\frac{X-\mu}{\lambda^2} f(\frac{X-\mu}{\lambda}).$$

Hence, the elements of the information gain matrix are given by

$$\Delta_{11} = \frac{1}{\lambda^2} E\left\{\frac{[f(X)]^2}{F(X)(1-F(X))}\right\},$$
$$\Delta_{22} = \frac{1}{\lambda^2} E\left\{\frac{[Xf(X)]^2}{F(X)(1-F(X))}\right\},$$
$$\Delta_{12} = \frac{1}{\lambda^2} E\left\{\frac{X[f(X)]^2}{F(X)(1-F(X))}\right\},$$

where the expectation is taken with respect to X which has distribution function $F(x)$ and density function $f(x)$. If $f(x)$ is symmetric, then $\Delta_{12} = 0$.

It is interesting to note that the information gain matrix is independent of the location parameter and inversely proportional to the squared scale parameter.

In particular, for normal family $N(\mu, \lambda^2)$, we have

$$\Delta_{11} = 0.4805/\lambda^2,$$
$$\Delta_{22} = 0.0675/\lambda^2,$$
$$\Delta_{12} = 0,$$

and, for exponential family $\mathcal{E}(\theta)$, we have

$$\Delta(\theta) = 0.4041/\theta^2.$$

Shape-scale family. Let $F(x;\theta) = F_0(\frac{x}{\lambda}, \alpha)$, where $\theta = (\alpha, \lambda)^T$. Let $F_\alpha(x) = F_0(x, \alpha)$ and $f_\alpha(x)$ denote its density function. For this distribution family, the information gain matrix is given by

$$\begin{pmatrix} 1 & 0 \\ 0 & \lambda^{-1} \end{pmatrix} \begin{pmatrix} \delta_{11}(\alpha) & \delta_{12}(\alpha) \\ \delta_{12}(\alpha) & \delta_{22}(\alpha) \end{pmatrix} \begin{pmatrix} 1 & 0 \\ 0 & \lambda^{-1} \end{pmatrix}$$

where

$$\delta_{11}(\alpha) = E\left\{\frac{[\partial F_\alpha(X)/\partial\alpha]^2}{F_\alpha(X)[1-F_\alpha(X)]}\right\},$$

$$\delta_{12}(\alpha) = -E\left\{\frac{X f_\alpha(X)\partial F_\alpha(X)/\partial\alpha}{F_\alpha(X)[1-F_\alpha(X)]}\right\},$$

$$\delta_{22}(\alpha) = E\left\{\frac{[X f_\alpha(X)]^2}{F_\alpha(X)[1-F_\alpha(X)]}\right\},$$

where the expectation is taken with respect to X which has distribution function $F_\alpha(x)$ and density function $f_\alpha(x)$.

3.1.2 The Fisher information of ranked set samples when ranking is imperfect*

In practice, only in rare cases, ranking in RSS could be perfect. In general, imperfect ranking is unavoidable. Imperfect ranking can arise due to: (i) ranking errors when judgment-ranking with respect to the variable of interest itself and (ii) discordance of the ranks and the real numerical orders when the ranks are induced by some other mechanism such as ranking with respect to certain concomitant variables. We investigate in this subsection whether a ranked set sample, while ranking is imperfect, still contains more information than a simple random sample in the Fisherian sense. We assume that the ranking mechanism is consistent. Further, assume that, for each $F_{[r]}$, the corresponding density function, denoted by $f_{[r]}$, exists, and that they satisfy the same regularity conditions assumed for $f_{(r)}$'s.

We have the following theorems.

Theorem 3.3. *Suppose the ranking mechanism in the RSS is consistent. Then, under the regularity conditions,*

$$I_{RSS}(\boldsymbol{\theta}, n) \geq nI(\boldsymbol{\theta}). \tag{3.5}$$

Theorem 3.4. *In addition to the assumptions made in Theorem 3.3, assume that for any non-zero* $\mathbf{a}$ *and any fixed* $\boldsymbol{\theta}$, *the distribution family* $f(\boldsymbol{x}, \eta) = f(\boldsymbol{x}, \boldsymbol{\theta} + \eta\mathbf{a})$ *is complete. Then,*

(1) The equality holds in (3.5) *for all* $\boldsymbol{\theta}$ *if and only if for almost all* x, $[F]$,

$$f_{[1]}(x;\boldsymbol{\theta}) = \cdots = f_{[k]}(x;\boldsymbol{\theta}).$$

(2) The information gain matrix $I_{RSS}(\boldsymbol{\theta}, n) - nI(\boldsymbol{\theta})$ *is not of full rank, if and only if there exists a vector* $\boldsymbol{b} \neq 0$, *such that for almost all* x, $[F]$,

$$\boldsymbol{b}'\boldsymbol{D}\log f_{[1]}(x;\boldsymbol{\theta}) = \cdots = \boldsymbol{b}'\boldsymbol{D}\log f_{[k]}(\boldsymbol{x};\boldsymbol{\theta}).$$

(3) The inequality in (3.5) is strict for a given $\boldsymbol{\theta}$ if and only if for each $\mathbf{b} \neq 0$ there is at least one r such that

$$P_{[r]}\Big(\boldsymbol{x};\ \boldsymbol{b}'\boldsymbol{D}\log f_{[r]}(\boldsymbol{x};\boldsymbol{\theta}) \neq \boldsymbol{b}'\boldsymbol{D}\log f(\boldsymbol{x};\boldsymbol{\theta})\Big) > 0.$$

The following lemma is needed in the proof of Theorem 3.3.

Lemma 3.5. *Let $\boldsymbol{x}_i$, $i = 1, \cdots, n$, be n vectors of dimension q, and a_i, $i = 1, \cdots, n$, be n positive numbers satisfying $\sum a_i = 1$. Then*

$$\sum_{i=1}^{n} a_i \boldsymbol{x}_i \boldsymbol{x}_i' \geq \Big(\sum_{i=1}^{n} a_i \boldsymbol{x}_i\Big)\Big(\sum_{i=1}^{n} a_i \boldsymbol{x}_i\Big)'.$$

Furthermore, if there does not exist a q-vector $\mathbf{b} \neq 0$ such that $\boldsymbol{b}'\boldsymbol{x}_i$ is independent of i, then

$$\sum_{i=1}^{n} a_i \boldsymbol{x}_i \boldsymbol{x}_i' > \Big(\sum_{i=1}^{n} a_i \boldsymbol{x}_i\Big)\Big(\sum_{i=1}^{n} a_i \boldsymbol{x}_i\Big)',$$

where $A > B$ means that $A - B$ is positive definite.

Proof. For any nonzero q-vector $\boldsymbol{b}$, define a random variable W by $P(W = \boldsymbol{b}'\boldsymbol{x}_i) = a_i$. Then, it follows from Cauchy-Schwarz inequality that

$$\boldsymbol{b}'\Big(\sum_{i=1}^{n} a_i \boldsymbol{x}_i \boldsymbol{x}_i'\Big)\boldsymbol{b} = \sum_{i=1}^{n} a_i (\boldsymbol{b}'\boldsymbol{x}_i)^2 = E(W^2)$$

$$\geq (E(W))^2 = \Big(\sum_{i=1}^{n} a_i (\boldsymbol{b}'\boldsymbol{x}_i)\Big)^2 = \boldsymbol{b}'\Big(\sum_{i=1}^{n} a_i \boldsymbol{x}_i\Big)\Big(\sum_{i=1}^{n} a_i \boldsymbol{x}_i\Big)'\boldsymbol{b}.$$

This inequality proves the first part of the lemma. Note that the equality sign holds in the above inequality if and only if $\boldsymbol{b}'\boldsymbol{x}_i$ is a constant in i. The second conclusion then follows.

Proof of Theorem 3.3: Under the regularity conditions, the Fisher information matrix of $\boldsymbol{X}_n$ exists. Hence, we have

$$\begin{aligned}
&I_{\text{RSS}}(\boldsymbol{\theta}, n)\\
&= -E\nabla \log(L(\boldsymbol{X}_n, \boldsymbol{\theta}))\\
&= -\sum_{r=1}^{k}\sum_{j=1}^{m} E\nabla \log(f_{[r]}(\boldsymbol{X}_{[r]j}, \boldsymbol{\theta}))\\
&= -m\sum_{r=1}^{k} E f_{[r]}^{-1}(\boldsymbol{X}_{[r]}; \boldsymbol{\theta})\nabla f_{[r]}(\boldsymbol{X}_{[r]}, \boldsymbol{\theta})\\
&+ m\sum_{r=1}^{k}\int \Big(\frac{\boldsymbol{D}f_{[r]}(\boldsymbol{x};\boldsymbol{\theta})}{f_{[r]}(\boldsymbol{x};\boldsymbol{\theta})}\Big)\Big(\frac{\boldsymbol{D}f_{[r]}(\boldsymbol{x};\boldsymbol{\theta})}{f_{[r]}(\boldsymbol{x};\boldsymbol{\theta})}\Big)^T f_{[r]}(\boldsymbol{x};\boldsymbol{\theta})d\boldsymbol{x}
\end{aligned}$$

$$= -m\sum_{r=1}^{k}\int \nabla f_{[r]}(\boldsymbol{x},\boldsymbol{\theta})d\boldsymbol{x}$$

$$+mk\int\sum_{r=1}^{k}\Big(\frac{\boldsymbol{D}f_{[r]}(\boldsymbol{x};\boldsymbol{\theta})}{f_{[r]}(\boldsymbol{x};\boldsymbol{\theta})}\Big)\Big(\frac{\boldsymbol{D}f_{[r]}(\boldsymbol{x};\boldsymbol{\theta})}{f_{[r]}(\boldsymbol{x};\boldsymbol{\theta})}\Big)^T\frac{f_{[r]}(\boldsymbol{x};\boldsymbol{\theta})}{kf(\boldsymbol{x};\boldsymbol{\theta})}f(\boldsymbol{x};\boldsymbol{\theta})d\boldsymbol{x}$$

$$= -mk\int \nabla f(\boldsymbol{x},\boldsymbol{\theta})d\boldsymbol{x}$$

$$+mk\int\sum_{r=1}^{k}\Big(\frac{\boldsymbol{D}f_{[r]}(\boldsymbol{x};\boldsymbol{\theta})}{f_{[r]}(\boldsymbol{x};\boldsymbol{\theta})}\Big)\Big(\frac{\boldsymbol{D}f_{[r]}(\boldsymbol{x};\boldsymbol{\theta})}{f_{[r]}(\boldsymbol{x};\boldsymbol{\theta})}\Big)^T\frac{f_{[r]}(\boldsymbol{x};\boldsymbol{\theta})}{kf(\boldsymbol{x};\boldsymbol{\theta})}f(\boldsymbol{x};\boldsymbol{\theta})d\boldsymbol{x}$$

$$\geq mk\int\sum_{r=1}^{k}\frac{\boldsymbol{D}f_{[r]}(\boldsymbol{x};\boldsymbol{\theta})}{f_{[r]}(\boldsymbol{x};\boldsymbol{\theta})}\frac{f_{[r]}(\boldsymbol{x};\boldsymbol{\theta})}{kf(\boldsymbol{x};\boldsymbol{\theta})}\sum_{r=1}^{k}\Big(\frac{\boldsymbol{D}f_{[r]}(\boldsymbol{x};\boldsymbol{\theta})}{f_{[r]}(\boldsymbol{x};\boldsymbol{\theta})}\frac{f_{[r]}(\boldsymbol{x};\boldsymbol{\theta})}{kf(\boldsymbol{x};\boldsymbol{\theta})}\Big)^T f(\boldsymbol{x};\boldsymbol{\theta})d\boldsymbol{x}$$

$$= mk\int\Big(\frac{1}{k}\sum_{r=1}^{k}\boldsymbol{D}f_{[r]}(\boldsymbol{x};\boldsymbol{\theta})\Big)\Big(\frac{1}{k}\sum_{r=1}^{k}\boldsymbol{D}f_{[r]}(\boldsymbol{x};\boldsymbol{\theta})\Big)^T f^{-1}(\boldsymbol{x};\boldsymbol{\theta})d\boldsymbol{x}$$

$$= mk\int\Big(\boldsymbol{D}f(\boldsymbol{x};\boldsymbol{\theta})\Big)\Big(\boldsymbol{D}f(\boldsymbol{x};\boldsymbol{\theta})\Big)^T f^{-1}(\boldsymbol{x};\boldsymbol{\theta})d\boldsymbol{x}$$

$$= mkI(\boldsymbol{\theta}),$$

where the last inequality follows from Lemma 3.5 and the final step follows from the fundamental equality. This completes the proof of Theorem 3.3.

Proof of Theorem 3.4: From the proof of Theorem 3.3, we can see that the information gain matrix is given by

$$\Delta(\boldsymbol{\theta}) = \frac{1}{k-1}\sum_{r=1}^{k}\int\Big(\frac{\boldsymbol{D}f_{[r]}(\boldsymbol{x})}{f_{[r]}(\boldsymbol{x})}-\frac{\boldsymbol{D}f(\boldsymbol{x})}{f(\boldsymbol{x})}\Big)\Big(\frac{\boldsymbol{D}f_{[r]}(\boldsymbol{x})}{f_{[r]}(\boldsymbol{x})}-\frac{\boldsymbol{D}f(\boldsymbol{x})}{f(\boldsymbol{x})}\Big)^T f_{[r]}(\boldsymbol{x})d\boldsymbol{x}.$$

Thus, part (1) of the theorem is true (*i.e.*, $\Delta = 0$) if and only if for every r,

$$\boldsymbol{D}\log f_{[r]}(\boldsymbol{x},\boldsymbol{\theta}) = \boldsymbol{D}\log f(\boldsymbol{x};\boldsymbol{\theta}), \quad \text{for almost all } \boldsymbol{x}[F_{[r]}].$$

These equations are equivalent to that $f_{[r]}(\boldsymbol{x},\boldsymbol{\theta}) = c_r(\boldsymbol{x})f(\boldsymbol{x},\boldsymbol{\theta})$ for all $\boldsymbol{\theta}$. The latter are, in turn, equivalent to $c_r(\boldsymbol{x}) \equiv 1$ which follows from the completeness of $f(\boldsymbol{x};\boldsymbol{\theta})$ and the fact that $\int c_r(\boldsymbol{x})f(\boldsymbol{x};\boldsymbol{\theta})d\boldsymbol{x} = 1$. Part (1) is proved.

As for part (2), we note that Δ is not of full rank if and only if there is a vector $\boldsymbol{b} \neq 0$ such that $\boldsymbol{b}^T\Delta\boldsymbol{b} = 0$. The latter is equivalent to that, for every r,

$$\boldsymbol{D}\log f_{[r]}(\boldsymbol{x};\boldsymbol{\theta}) = \boldsymbol{D}\log f(\boldsymbol{x};\boldsymbol{\theta}) \quad a.e.\boldsymbol{x}[F_{[r]}].$$

Finally, it follows from Lemma 3.5 that the strict inequality in the proof of Theorem 3.3 holds if and only if, for every vector $\boldsymbol{b} \neq 0$, there is at least one r such that

$$P_{[r]}\Big(\boldsymbol{b}'\boldsymbol{D}\log f_{[r]}(x;\boldsymbol{\theta}) \neq \boldsymbol{b}'\boldsymbol{D}\log f(\boldsymbol{x};\boldsymbol{\theta})\Big) > 0.$$

The above theorems imply that, as long as the ranking in RSS is not done at random, the RSS always provides more information than SRS even if ranking is imperfect. In the remainder of this subsection, we shall derive a result which tells us how much information gain is obtained and how the information gain depends on the ranking errors. Towards this end, we first derive a lemma which expresses $F_{[r]}$'s in terms of $F_{(r)}$'s.

Lemma 3.6. *Suppose that the ranking mechanism in the RSS is consistent. Then, $F_{[r]}$ can be expressed as*

$$F_{[r]} = \sum_{s=1}^{k} p_{rs} F_{(s)},$$

where $p_{rs} \geq 0$ and $\sum_{s=1}^{k} p_{rs} = \sum_{r=1}^{k} p_{rs} = 1$.

Proof. Let $p_{rs} = P(X_{[r]} = X_{(s)})$, that is, p_{rs} is the probability with which a unit receives rank r but indeed its X is the sth order statistic. It is trivial that $\sum_{s=1}^{k} p_{rs} = 1$. Therefore, the distribution of $X_{[r]}$ can be expressed as a mixture of the distributions of $X_{(s)}, s = 1, \ldots, k$ with mixture weights p_{rs}. Furthermore, since both $F_{[r]}$'s and $F_{(r)}$'s satisfy the fundamental equality, we have

$$\sum_{s=1}^{k} (\sum_{r=1}^{k} p_{rs} - 1) F_{(s)}(x) = 0$$

for all x, which implies that $\sum_{r=1}^{k} p_{rs} = 1$.

By using Lemma 3.6, $f_{[r]}$ can be expressed as

$$\begin{aligned} f_{[r]}(x, \boldsymbol{\theta}) &= \sum_{s=1}^{k} p_{sr} \frac{k!}{(k-s)!(s-1)!} F^{s-1}(x, \boldsymbol{\theta})[1 - F(x, \boldsymbol{\theta})]^{k-s} f(x, \boldsymbol{\theta}) \\ &= g_r(x, \boldsymbol{\theta}) f(x, \boldsymbol{\theta}), \quad \text{say.} \end{aligned}$$

Since

$$\sum_{r=1}^{k} f_{[r]}(x, \boldsymbol{\theta}) = k f(x, \boldsymbol{\theta}),, \tag{3.6}$$

we have

$$\sum_{r=1}^{k} g_r(x, \boldsymbol{\theta}) = k. \tag{3.7}$$

We now state the following result.

Theorem 3.7. *Suppose that the ranking mechanism in the RSS is consistent. Then, under the regularity conditions,*

$$I_{RSS}(\boldsymbol{\theta}, n) = mkI(\boldsymbol{\theta}) + \tilde{\Delta}(\boldsymbol{\theta}),$$

where $\tilde{\Delta}(\boldsymbol{\theta})$ is a non-negative definite matrix given by

$$\tilde{\Delta}(\boldsymbol{\theta}) = \sum_{r=1}^{k} E\left[\frac{\boldsymbol{D}g_r(X,\boldsymbol{\theta})(\boldsymbol{D}g_r(X,\boldsymbol{\theta}))^T}{g_r(X,\boldsymbol{\theta})}\right].$$

Here the expectation is taken with respect to $X(\sim F)$.

Proof. It suffices to prove the above result in the case $m = 1$. We have

$$\begin{aligned} &I_{\text{RSS}}(\boldsymbol{\theta}, n) \\ &= -E\nabla \sum_{r=1}^{k} \log(f_{[r]}(X_{[r]}, \boldsymbol{\theta})) \\ &= -E\nabla \sum_{r=1}^{k} \log(g_{[r]}(X_{[r]}, \boldsymbol{\theta})) - E\nabla \sum_{r=1}^{k} \log(f(X_{[r]}, \boldsymbol{\theta})). \end{aligned} \tag{3.8}$$

It follows from (3.6) that

$$-E\nabla \sum_{r=1}^{k} \log(f(X_{[r]}, \boldsymbol{\theta})) = kI(\theta). \tag{3.9}$$

Write

$$\begin{aligned} &-E\nabla \sum_{r=1}^{k} \log(g_{[r]}(X_{[r]}, \boldsymbol{\theta})) \\ &= \sum_{r=1}^{k} E\left[\frac{\boldsymbol{D}g_r(X_{[r]},\boldsymbol{\theta})(\boldsymbol{D}g_r(X_{[r]},\boldsymbol{\theta}))^T}{g_r(X_{[r]},\boldsymbol{\theta})^2}\right] - \sum_{r=1}^{k} E\left[\frac{\nabla g_r(X_{[r]},\boldsymbol{\theta})}{g_r(X_{[r]},\boldsymbol{\theta})}\right]. \end{aligned} \tag{3.10}$$

It follows from (3.7) that

$$\sum_{r=1}^{k} E\left[\frac{\nabla g_r(X_{[r]},\boldsymbol{\theta})}{g_r(X_{[r]},\boldsymbol{\theta})}\right] = \int \nabla\left(\sum_{r=1}^{k} g_r(x,\boldsymbol{\theta})\right) f(x,\boldsymbol{\theta})dx = 0. \tag{3.11}$$

Finally, we have

$$\begin{aligned} &\sum_{r=1}^{k} E\left[\frac{\boldsymbol{D}g_r(X_{[r]},\boldsymbol{\theta})(\boldsymbol{D}g_r(X_{[r]},\boldsymbol{\theta}))^T}{g_r(X_{[r]},\boldsymbol{\theta})^2}\right] \\ &= \sum_{r=1}^{k} \int \boldsymbol{D}g_r(x,\boldsymbol{\theta})(\boldsymbol{D}g_r(x,\boldsymbol{\theta}))^T \frac{f(x,\boldsymbol{\theta})}{g_r(x,\boldsymbol{\theta})} dx \\ &= \sum_{r=1}^{k} E\left[\frac{\boldsymbol{D}g_r(X,\boldsymbol{\theta})(\boldsymbol{D}g_r(X,\boldsymbol{\theta}))^T}{g_r(X,\boldsymbol{\theta})}\right]. \end{aligned} \tag{3.12}$$

The theorem then follows from (3.8) — (3.12).

3.2 The maximum likelihood estimate and its asymptotic relative efficiency

In this section, we deal with the MLE of $\boldsymbol{\theta}$ using a ranked set sample under perfect ranking and consider its asymptotic relative efficiency with respect to the MLE under a simple random sample.

Let $\hat{\boldsymbol{\theta}}_{n\text{SRS}}$ denote the MLE of $\boldsymbol{\theta}$ using a simple random sample of size n. It is well-known that, under certain regularity conditions, $\hat{\boldsymbol{\theta}}_{n\text{SRS}}$ is strongly consistent and asymptotically normal with mean $\boldsymbol{\theta}$ and variance-covariance matrix $I^{-1}(\boldsymbol{\theta})/n$.

Let $\hat{\boldsymbol{\theta}}_{n\text{RSS}}$ denote the MLE of $\boldsymbol{\theta}$ using a ranked set sample of size $n = mk$. We have similar results for the RSS version of the MLE.

Theorem 3.8. *Under certain regularity conditions,*

(i) $\hat{\boldsymbol{\theta}}_{nRSS}$ *is strongly consistent as n goes to infinity.*
(ii) As n goes to infinity,

$$\sqrt{n}(\hat{\boldsymbol{\theta}}_{nRSS} - \boldsymbol{\theta}) \to N(0, \Sigma_{RSS}),$$

in distribution, where

$$\Sigma_{RSS} = [I(\boldsymbol{\theta}) + (k-1)\Delta(\boldsymbol{\theta})]^{-1}.$$

The proof of Theorem 3.8 can be carried out in the same way as in simple random sampling case and hence is omitted.

If $\phi = \phi(\boldsymbol{\theta})$ is a function of $\boldsymbol{\theta}$ then, by the functional invariance of MLE, the MLE of ϕ is given by $\phi(\hat{\boldsymbol{\theta}}_{N\text{SRS}})$ and $\phi(\hat{\boldsymbol{\theta}}_{N\text{RSS}})$, respectively, in the RSS and the SRS. If ϕ is differentiable then both estimators are strongly consistent and asymptotically normal with asymptotic variances given, respectively, by

$$\sigma^2_{\phi\text{SRS}} = \left[\frac{\partial\phi(\boldsymbol{\theta})}{\partial\boldsymbol{\theta}}\right]^T [I(\theta)]^{-1}\frac{\partial\phi(\boldsymbol{\theta})}{\partial\boldsymbol{\theta}}$$

and

$$\sigma^2_{\phi\text{RSS}} = \left[\frac{\partial\phi(\boldsymbol{\theta})}{\partial\boldsymbol{\theta}}\right]^T [I(\theta) + (k-1)\Delta(\theta)]^{-1}\frac{\partial\phi(\boldsymbol{\theta})}{\partial\boldsymbol{\theta}}.$$

The ARE of $\phi(\hat{\boldsymbol{\theta}}_{N\text{RSS}})$ with respect to $\phi(\hat{\boldsymbol{\theta}}_{N\text{SRS}})$ is then given by

$$\text{ARE}(\phi(\hat{\boldsymbol{\theta}}_{N\text{SRS}}), \phi(\hat{\boldsymbol{\theta}}_{N\text{SRS}})) = \frac{\sigma^2_{\phi\text{RSS}}}{\sigma^2_{\phi\text{RSS}}}.$$

In particular, if $\boldsymbol{\theta}$ itself is a scalar, the ARE becomes

$$\text{ARE}(\phi(\hat{\boldsymbol{\theta}}_{N\text{SRS}}), \phi(\hat{\boldsymbol{\theta}}_{N\text{SRS}})) = 1 + (k-1)\frac{\Delta(\theta)}{I(\theta)}.$$

Note that since the matrix $\Delta(\theta)$ is non-negative definite the ARE is always greater than or equal to 1. It implies that the MLE based on a ranked set sample is always more efficient than the MLE based on a simple random sample.

In what follows, we examine the ARE in the estimation of the mean and variance for some particular families.

(a) Normal family $N(\mu, \sigma^2)$. The information matrix about $\boldsymbol{\theta} = (\mu, \sigma)$ of a single random observation from a normal distribution is as follows:

$$I(\mu, \sigma) = \frac{1}{\sigma^2}\begin{pmatrix} 1 & 0 \\ 0 & 4 \end{pmatrix}.$$

As given in Section 3.1.1, the information gain matrix for the Normal family is

$$\Delta(\mu, \sigma) = \frac{1}{\sigma^2}\begin{pmatrix} 0.4805 & 0 \\ 0 & 0.0675 \end{pmatrix}.$$

Thus, we obtain that the AREs for the estimation of μ and σ are, respectively, $1 + 0.4805(k-1)$ and $1 + 0.0169(k-1)$.

Table 3.1. The ARE of the parametric RSS with $k = 2$ in the estimation of mean and variance, the skewness γ and the kurtosis κ for selected Gamma distributions

α	ARE(μ)	ARE(σ^2)	γ	κ
1.5	1.4192	1.2321	1.6330	7.0000
2	1.4302	1.1979	1.4142	6.0000
2.5	1.4409	1.2072	1.2649	5.4000
3	1.4458	1.1854	1.1547	5.0000
3.5	1.4504	1.1807	1.0690	4.7143
4	1.4540	1.1767	1.0000	4.5000
4.5	1.4568	1.1733	0.9428	4.3333
5	1.4590	1.1702	0.8944	4.2000
5.5	1.4641	1.1700	0.8528	4.0909
6	1.4623	1.1659	0.8165	4.0000
6.5	1.4637	1.1641	0.7845	3.9231
7	1.4649	1.1624	0.7559	3.8571
7.5	1.4660	1.1613	0.7303	3.8000
8	1.4670	1.1595	0.7071	3.7500
8.5	1.4679	1.1584	0.6860	3.7059
9	1.4686	1.1556	0.6667	3.6667
9.5	1.4688	1.1545	0.6489	3.6316
10	1.4687	1.1528	0.6325	3.6000

(b). Gamma family $Gamma(\alpha, \lambda)$. The information matrix $I(\alpha, \lambda)$ is as follows:

$$\begin{pmatrix} [\Gamma(\alpha)\Gamma''(\alpha) - [\Gamma'(\alpha)]^2][\Gamma(\alpha)]^{-2} & \lambda^{-1} \\ \lambda^{-1} & \alpha\lambda^{-2} \end{pmatrix}.$$

Table 3.2. The ARE of the parametric RSS with $k = 2$ in the estimation of mean and variance, the skewness γ and the kurtosis κ for selected Weibull distributions

α	ARE(μ)	ARE(σ^2)	γ	κ
1.5	1.44281	1.15844	1.0720	4.3904
2	1.46693	1.14179	0.6311	3.2451
2.5	1.47771	1.13122	0.3586	2.8568
3	1.48277	1.12695	0.1681	2.7295
3.5	1.48517	1.12676	0.0251	2.7127
4	1.48626	1.12876	-0.0872	2.7478
4.5	1.48667	1.13175	-0.1784	2.8081
5	1.48670	1.13509	-0.2541	2.8803
5.5	1.48651	1.13845	-0.3182	2.9574
6	1.48619	1.14166	-0.3733	3.0355
6.5	1.48581	1.14465	-0.4211	3.1125
7	1.48539	1.14739	-0.4632	3.1872
7.5	1.48498	1.14991	-0.5005	3.2590
8	1.48458	1.15222	-0.5337	3.3277
8.5	1.48417	1.15432	-0.5636	3.3931
9	1.48375	1.15623	-0.5907	3.4552
9.5	1.48343	1.15809	-0.6152	3.5142
10	1.48316	1.15978	-0.6376	3.5702

The mean and the variance of the Gamma distribution, as functions of α and λ, are given, respectively, by $\mu = \phi_1 = \alpha\lambda$ and $\sigma^2 = \phi_2 = \alpha\lambda^2$. We have $\phi_1' = (\lambda, \alpha)^T$ and $\phi_2' = (\lambda^2, 2\alpha\lambda)^T$. The ARE for the estimation of the mean and variance can be calculated by using the information gain matrix given in Section 3.1.1 and the above expressions for $I(\alpha, \lambda)$, ϕ_1' and ϕ_2'.

(c). Weibull family $Weibull(\lambda, \alpha)$. The information matrix $I(\alpha, \lambda)$ is as follows:

$$\begin{pmatrix} \alpha^{-2} + \tau_2 & \lambda^{-1}(1 - \alpha\tau_1 - \tau_0) \\ \lambda^{-1}(1 - \alpha\tau_1 - \tau_0) & \alpha\lambda^{-2}[(\alpha + 1)\tau_0 - 1] \end{pmatrix},$$

where $\tau_i = E[X^\alpha(lnX)^i], i = 0, 1, 2$, with $X \sim Weibull(\alpha, 1)$. The mean and the variance of the Weibull distribution, as functions of α and λ, are given by $\mu = \phi_1 = \lambda\Gamma(1 + 1/\alpha)$ and $\sigma^2 = \phi_2 = \lambda^2[\Gamma(1 + 2/\alpha) - \Gamma^2(1 + 1/\alpha)]$, respectively. We have

$$\phi_1' = [-\frac{\lambda}{\alpha^2}\Gamma'(1 + \frac{1}{\alpha}),\ \Gamma(1 + \frac{1}{\alpha})]^T$$

$$\phi_2' = [-\frac{2\lambda^2}{\alpha^2}(\Gamma'(1 + \frac{2}{\alpha}) - \Gamma(1 + \frac{1}{\alpha})\Gamma'(1 + \frac{1}{\alpha})),$$
$$2\lambda(\Gamma(1 + \frac{2}{\alpha}) - \Gamma^2(1 + \frac{1}{\alpha}))]^T.$$

Similarly, the ARE for the estimation of mean and variance can be calculated by using the information gain matrix given in section 3.1.1 and the expressions above.

The AREs of the RSS with $k = 2$ for the estimation of the mean and variance for the Gamma and Weibull families with $\lambda = 1$ and some selected α values are given, respectively, in Table 3.1 and Table 3.2. These two tables can be compared with Table 2.1. The ARE of the MLE of the mean is always larger than the ARE of the sample mean for these two families. This is also true for other distributions. This seems to be a general phenomenon, and hence we might conclude that, in general, more gain can be obtained in parametric RSS than in nonparametric RSS.

We can also get a rough picture from Tables 3.1 and 3.2 on how the skewness or kurtosis affects the ARE. Since the skewness and kurtosis given in the two tables are highly positively correlated (for the Gamma family, the correlation is 0.9910, and for the Weibull family, the correlation is 0.9637), the effects of the skewness and the kurtosis on the ARE are aliased and can not be distinguished. We can observe the following features: the ARE in the estimation of the mean decreases as the skewness (or the kurtosis) increases, and the ARE in the estimation of the variance increases as the skewness (or the kurtosis) increases.

3.3 The best linear unbiased estimation for location-scale families

If the underlying distribution belongs to a location-scale family, the best linear unbiased estimation (BLUE) provides an alternative to the MLE for the estimation of the parameters. In the following, the BLUE and its properties are discussed.

If X follows the location-scale distribution $F(\frac{x-\mu}{\sigma})$ then $Z = \frac{X-\mu}{\sigma}$ follows the distribution $F(z)$. The order statistic $Z_{(r)i}$ is related to the order statistic $X_{(r)i}$ by $Z_{(r)i} = \frac{X_{(r)i}-\mu}{\sigma}$. Let

$$\alpha_r = E[Z_{(r)i}] \quad \nu_r = Var[Z_{(r)i}].$$

We can express the $X_{(r)i}$'s in the form of a linear regression model as follows:

$$X_{(r)i} = \mu + \sigma\alpha_r + \epsilon_{ri}, \quad r = 1, \ldots, k; i = 1, \ldots, m,$$

where ϵ_{ri} are independent random variables with $E\epsilon_{ri} = 0$ and $\mathrm{Var}(\epsilon_{ri}) = \sigma^2\nu_r$.

Let

$$\boldsymbol{\beta} = (\mu, \sigma)^T,$$

$$\boldsymbol{X} = (X_{(1)1}, \ldots, X_{(1)m}, \ldots, X_{(k)1}, \ldots, X_{(k)m})^T,$$

$$U = \begin{bmatrix} \mathbf{1} & \alpha_1 \mathbf{1} \\ \vdots & \vdots \\ \mathbf{1} & \alpha_k \mathbf{1} \end{bmatrix},$$

$$W = \begin{bmatrix} \nu_1 I & & \\ & \ddots & \\ & & \nu_k I \end{bmatrix},$$

where $\mathbf{1}$ is an m-dimensional vector whose elements are all 1 and I is an $m \times m$ identity matrix. By the theory of linear regression analysis, the BLUE of $\boldsymbol{\beta}$ is given by

$$\hat{\boldsymbol{\beta}} = (U^T W^{-1} U)^{-1} U^T W^{-1} \boldsymbol{X}.$$

The variance of $\hat{\boldsymbol{\beta}}$ is given by

$$\mathrm{Var}(\hat{\boldsymbol{\beta}}) = \sigma^2 (U^T W^{-1} U)^{-1}.$$

We also have that, as $m \to \infty$,

$$\sqrt{mk}(\hat{\boldsymbol{\beta}} - \boldsymbol{\beta}) \to N(0, \sigma^2 V^{-1}),$$

in distribution, where

$$V = \begin{pmatrix} \frac{1}{k}\sum_{r=1}^k \frac{1}{\nu_r} & \frac{1}{k}\sum_{r=1}^k \frac{\alpha_r}{\nu_r} \\ \frac{1}{k}\sum_{r=1}^k \frac{\alpha_r}{\nu_r} & \frac{1}{k}\sum_{r=1}^k \frac{\alpha_r^2}{\nu_r} \end{pmatrix}.$$

From the matrix expressions above, we can obtain that the BLUE of μ and σ are given, respectively, by

$$\hat{\sigma}_{\mathrm{BLUE}} = \frac{\sum_{r=1}^k \nu_r^{-1}(\alpha_r - \bar{\alpha})\bar{X}_{(r)}}{\sum_{r=1}^k \nu_r^{-1}(\alpha_r - \bar{\alpha})^2},$$

$$\hat{\mu}_{\mathrm{BLUE}} = \bar{X}_W - \bar{\alpha}\hat{\sigma},$$

where

$$\bar{\alpha} = \frac{1}{k}\sum_{r=1}^k \alpha_r, \; \bar{X}_{(r)} = \frac{1}{m}\sum_{i=1}^m X_{(r)i}, \; \bar{X}_W = \frac{\sum_{r=1}^k \nu_r^{-1}\bar{X}_{(r)}}{\sum_{r=1}^k \nu_r^{-1}}.$$

Now consider the special case that the underlying location-scale family is symmetric. Because of symmetry, we have

$$\sum_{r=1}^k \alpha_r = 0 \text{ and } \sum_{r=1}^k \frac{\alpha_r}{\nu_r} = 0.$$

Hence the BLUEs of μ and σ reduce to

$$\hat{\mu}_{\text{BLUE}} = \bar{X}_W,$$
$$\hat{\sigma}_{\text{BLUE}} = \frac{\sum_{r=1}^{k} \nu_r^{-1} \alpha_r \bar{X}_{(r)}}{\sum_{r=1}^{k} \nu_r^{-1} \alpha_r^2}.$$

The matrix V reduces to

$$V = \begin{pmatrix} \frac{1}{k}\sum_{r=1}^{k} \frac{1}{\nu_r} & 0 \\ 0 & \frac{1}{k}\sum_{r=1}^{k} \frac{\alpha_r^2}{\nu_r} \end{pmatrix}.$$

Thus the estimates $\hat{\mu}_{\text{BLUE}}$ and $\hat{\sigma}_{\text{BLUE}}$ are independent.

We remark that, in general, the BLUE is not as efficient as the MLE. However, it is much easier to compute. The BLUE should only be used instead of the MLE if the maximization of the likelihood function $l(\boldsymbol{\theta})$ is too complicated to produce a reliable maximizer. We should be even more cautious when trying to use the BLUE of σ. Usually, a linear estimate of a variance or any dispersion parameter is not as good as a quadratic estimate. Moreover, a BLUE of σ may produce a negative estimate. If a reliable MLE of σ can not be obtained, the non-parametric estimate discussed in Section 2.4 might provide an estimate better than the BLUE.

3.4 The regularity conditions and the existence of Fisher information matrix of ranked set samples*

In the discussion of Fisher information matrix in Section 3.1, we deliberately avoided the details of the regularity conditions and the existence of the Fisher information matrix needed for the theorems of that section in order not to distract the attention of the reader from the main results. In this section, we give these regularity conditions and prove the existence of the Fisher information matrix for mathematical rigor. Here we point out that, when these regularity conditions are satisfied, the MLE based on an RSS sample has the usual optimality properties of MLE, since the distribution of the ranked statistics is just a special case of parametric families.

The regularity conditions necessary for the theorems are as follows:

1. The support of $f(x;\boldsymbol{\theta})$ does not depend on $\boldsymbol{\theta}$.
2. For almost all $x[f(x;\boldsymbol{\theta})]$, the density functions are twice continuously differentiable with respect to $\boldsymbol{\theta}$, where the term "almost all $x[f(x;\boldsymbol{\theta})]$" is the abbreviation for "almost all x with respect to the measure generated by $f(x;\boldsymbol{\theta})$".
3. For each $\boldsymbol{\theta}_0 \in \Theta$, there is a neighborhood $N(\boldsymbol{\theta}_0) \subset \Theta$ and an integrable function $h(x)$ such that for any $i, j \le q$ and $\boldsymbol{\theta} \in N(\boldsymbol{\theta}_0)$,

$$\left|\frac{\partial f(x;\boldsymbol{\theta})}{\partial \theta_j}\right| \le h(x) \quad \text{and} \quad \left|\frac{\partial^2 f(x;\boldsymbol{\theta})}{\partial \theta_i \partial \theta_j}\right| \le h(x).$$

4. For each $\boldsymbol{\theta}$, $I(\boldsymbol{\theta}) = \left[E\frac{\partial \log f(X,\boldsymbol{\theta})}{\partial \theta_i}\frac{\partial \log f(X,\boldsymbol{\theta})}{\partial \theta_j}\right]_{i,j=1}^{q}$ exists.

We now proceed to prove the existence of the Fisher information matrix. The existence is guaranteed by the following lemma.

Lemma 3.9. *Under the regularity conditions, the Fisher information $I_{(r)}(\boldsymbol{\theta})$ of $X_{(r)}$ exists and its (i,j)-th element is*

$$
\begin{aligned}
I_{(r)ij} &= \int \Big(D_i \log(f_{(r)}(x;\boldsymbol{\theta}))D_j \log(f_{(r)}(x;\boldsymbol{\theta}))\Big) f_{(r)}(x;\boldsymbol{\theta})dx \\
&= -\int \Big(\nabla_{ij} \log(f_{(r)}(x;\boldsymbol{\theta}))\Big) f_{(r)}(x;\boldsymbol{\theta})dx.
\end{aligned}
$$

Proof. To prove the lemma, we only need to verify the regularity conditions for $f_{(r)}(x;\boldsymbol{\theta})$. Note that

$$
D_i f_{(r)}(x;\boldsymbol{\theta}) = f_{(r)}(x;\boldsymbol{\theta})[(r-1)\frac{F_i'(x;\boldsymbol{\theta})}{F(x;\boldsymbol{\theta})} - (k-r)\frac{F_i'(x;\boldsymbol{\theta})}{1-F(x;\boldsymbol{\theta})} + \frac{f_i'(x;\boldsymbol{\theta})}{f(x;\boldsymbol{\theta})}]
$$

and

$$
\begin{aligned}
&\nabla_{ij} f_{(r)}(x;\boldsymbol{\theta}) \\
&= f_{(r)}(x;\boldsymbol{\theta})[(r-1)\frac{F_{ij}''(x;\boldsymbol{\theta})}{F(x;\boldsymbol{\theta})} - (k-r)\frac{F_{ij}''(x;\boldsymbol{\theta})}{1-F(x;\boldsymbol{\theta})} + \frac{f_{ij}''(x;\boldsymbol{\theta})}{f(x;\boldsymbol{\theta})}] \\
&+f_{(r)}(x;\boldsymbol{\theta})[(r-1)(r-2)\frac{F_i'(x;\boldsymbol{\theta})F_j'(x;\boldsymbol{\theta})}{F^2(x;\boldsymbol{\theta})} \\
&+(k-r)(k-r-1)\frac{F_i'(x;\boldsymbol{\theta})F_j'(x;\boldsymbol{\theta})}{(1-F(x;\boldsymbol{\theta}))^2}] \\
&-2f_{(r)}(x;\boldsymbol{\theta})[(r-1)(k-r)\frac{F_i'(x;\boldsymbol{\theta})F_j'(x;\boldsymbol{\theta})}{F(x;\boldsymbol{\theta})(1-F(x;\boldsymbol{\theta}))} \\
&+2f_{(r)}(x;\boldsymbol{\theta})[(r-1)\frac{F_i'(x;\boldsymbol{\theta})}{F(x;\boldsymbol{\theta})} - (k-r)\frac{F_i'(x;\boldsymbol{\theta})}{1-F(x;\boldsymbol{\theta})}]\frac{f_j'(x;\boldsymbol{\theta})}{f(x;\boldsymbol{\theta})} \\
&+2f_{(r)}(x;\boldsymbol{\theta})[(r-1)\frac{F_j'(x;\boldsymbol{\theta})}{F(x;\boldsymbol{\theta})} - (k-r)\frac{F_j'(x;\boldsymbol{\theta})}{1-F(x;\boldsymbol{\theta})}]\frac{f_i'(x;\boldsymbol{\theta})}{f(x;\boldsymbol{\theta})}.
\end{aligned}
$$

From these one can easily see that, for almost all $x[f(x;\boldsymbol{\theta})]$, $f_{(r)}(x;\boldsymbol{\theta})$ is continuously twice differentiable.

Next, we need to show that $D_i f_{(r)}(x;\boldsymbol{\theta})$ is controlled by some integrable function $H(x)$ uniformly for $\boldsymbol{\theta} \in N(\boldsymbol{\theta}_0)$. Without loss of generality, we assume that the diameter of $N(\boldsymbol{\theta}_0)$ is less than 1. Write $f_{(r)}(x;\boldsymbol{\theta})$ by $f_{(r:k)}(x;\boldsymbol{\theta})$. We have

$$
\begin{aligned}
D_i f_{(r)}(x;\boldsymbol{\theta}) &= k(f_{(r-1:k-1)}(x;\boldsymbol{\theta}) - f_{(r:k-1)}(x;\boldsymbol{\theta}))F_i'(x;\boldsymbol{\theta}) \\
&+\frac{k!}{(r-1)!(k-r)!}F^{r-1}(x;\boldsymbol{\theta})(1-F(x;\boldsymbol{\theta}))^{k-r}f_i'(x;\boldsymbol{\theta}).
\end{aligned}
$$

Then,

$$\begin{aligned} f(x;\boldsymbol{\theta}) &= f(x;\boldsymbol{\theta}_0) + \int_0^1 (\boldsymbol{\theta}-\boldsymbol{\theta}_0)^T \boldsymbol{D} f(x;\boldsymbol{\theta}_0 + t(\boldsymbol{\theta}-\boldsymbol{\theta}_0))dt \\ &\leq f(x;\boldsymbol{\theta}_0) + \sqrt{q}h(x), \end{aligned}$$

which is integrable. By the regularity conditions, one can easily show that

$$F_i'(x;\boldsymbol{\theta}) = \int_{-\infty}^{x} f_i'(y;\boldsymbol{\theta})dy. \tag{3.13}$$

Let $\boldsymbol{\theta}_\eta$ be the vector obtained from $\boldsymbol{\theta}$ by adding η to its ith component. When η is small enough, in fact, the whole segment between $\boldsymbol{\theta}$ and $\boldsymbol{\theta}_\eta$ is in $N(\boldsymbol{\theta})$. Thus,

$$\begin{aligned} F_i'(x;\boldsymbol{\theta}) &= \lim_{\eta\to 0} \eta^{-1}(F(x;\boldsymbol{\theta}_\eta) - F(x;\boldsymbol{\theta})) \\ &= \lim_{\eta\to 0} \int_{-\infty}^{x} \eta^{-1}[f(y;\boldsymbol{\theta}_\eta) - f(y;\boldsymbol{\theta})]dy \\ &= \int_{-\infty}^{x} f_i'(y;\boldsymbol{\theta})dy, \end{aligned}$$

where the last step follows from Dominated Convergence Theorem and the fact that

$$|\eta^{-1}[f(y;\boldsymbol{\theta}_\eta) - f(y;\boldsymbol{\theta})]| = |f_i'(y;\boldsymbol{\theta}_*)| \leq h(y)$$

which is integrable, where $\boldsymbol{\theta}_*$ is a point in the segment connecting $\boldsymbol{\theta}_0$ and $\boldsymbol{\theta}_\eta$. Then,

$$|F_i'(x;\boldsymbol{\theta})| \leq \int_{-\infty}^{\infty} h(y)dy := C.$$

Therefore,

$$\left|D_i f_{(r)}(x;\boldsymbol{\theta})\right| \leq k![Cf(x;\boldsymbol{\theta}_0) + (C\sqrt{q}+1)h(x)].$$

By similar arguments, we have

$$\left|\nabla_{ij} f_{(r)}(x;\boldsymbol{\theta})\right| \leq k![(C+4C^2)f(x;\boldsymbol{\theta}_0) + ((C+4C^2)\sqrt{q}+3)h(x)].$$

Thus, $f_{(r)}(x;\boldsymbol{\theta})$ satisfies the second regularity condition with

$$H(x) = k![(C+4C^2)f(x;\boldsymbol{\theta}_0) + ((C+4C^2)\sqrt{q}+3)h(x)].$$

To prove the existence of the information matrix of $X_{(r)}$, we only need to prove the finiteness of its diagonal elements. We have

$$D_i \log f_{(r)}(x;\boldsymbol{\theta}) = (r-1)\frac{F_i'(x;\boldsymbol{\theta})}{F(x;\boldsymbol{\theta})} - (k-r)\frac{F_i'(x;\boldsymbol{\theta})}{1-F(x;\boldsymbol{\theta})} + \frac{f_i'(x;\boldsymbol{\theta})}{f(x;\boldsymbol{\theta})} \tag{3.14}$$

Thus, we need to show that

$$\int_{-\infty}^{\infty}\left(\frac{F_i'(x;\boldsymbol{\theta})}{F(x;\boldsymbol{\theta})}\right)^2 f_{(r)}(x;\boldsymbol{\theta})dx<\infty, r=2,\cdots,k, \tag{3.15}$$

$$\int_{-\infty}^{\infty}\left(\frac{F_i'(x;\boldsymbol{\theta})}{1-F(x;\boldsymbol{\theta})}\right)^2 f_{(r)}(x;\boldsymbol{\theta})dx<\infty, r=1,\cdots,k-1, \tag{3.16}$$

$$\int_{-\infty}^{\infty}\left(\frac{f_i'(x;\boldsymbol{\theta})}{f(x;\boldsymbol{\theta})}\right)^2 f_{(r)}(x;\boldsymbol{\theta})dx<\infty. \tag{3.17}$$

To prove (3.15), note from Cauchy-Schwarz inequality that

$$\left[\frac{F_i'(x;\boldsymbol{\theta})}{F(x;\boldsymbol{\theta})}\right]^2=\left[\int_{-\infty}^{x}\frac{f_i'(y;\boldsymbol{\theta})}{f(y;\boldsymbol{\theta})}\frac{f(y;\boldsymbol{\theta})}{F(x;\boldsymbol{\theta})}dy\right]^2\le\int_{-\infty}^{x}\left[\frac{f_i'(y;\boldsymbol{\theta})}{f(y;\boldsymbol{\theta})}\right]^2\frac{f(y;\boldsymbol{\theta})}{F(x;\boldsymbol{\theta})}dy.$$

Therefore, for $r=2,\cdots,k$,

$$\begin{aligned}
&\int_{-\infty}^{\infty}\left(\frac{F_i'(x;\boldsymbol{\theta})}{F(x;\boldsymbol{\theta})}\right)^2 f_{(r)}(x;\boldsymbol{\theta})dx\\
&=\int_{-\infty}^{\infty}\int_{-\infty}^{x}\left(\frac{f_i'(y;\boldsymbol{\theta})}{f(y;\boldsymbol{\theta})}\right)^2\frac{f(y;\boldsymbol{\theta})}{F(x;\boldsymbol{\theta})}dyf_{(r)}(x;\boldsymbol{\theta})dx\\
&\le\int_{-\infty}^{\infty}\left(\frac{f_i'(y;\boldsymbol{\theta})}{f(y;\boldsymbol{\theta})}\right)^2 f(y;\boldsymbol{\theta})dy\int_{-\infty}^{\infty}\frac{f_{(r)}(x;\boldsymbol{\theta})}{F(x;\boldsymbol{\theta})}dx\\
&=\frac{k}{r-1}\int_{-\infty}^{\infty}\left(\frac{f_i'(y;\boldsymbol{\theta})}{f(y;\boldsymbol{\theta})}\right)^2 f(y;\boldsymbol{\theta})dy<\infty.
\end{aligned}$$

Similarly, one can show that, for $r=1,\cdots,k-1$,

$$\int_{-\infty}^{\infty}\left[\frac{F_i'(x;\boldsymbol{\theta})}{1-F(x;\boldsymbol{\theta})}\right]^2 f_{(r)}(x;\boldsymbol{\theta})dx\le\frac{k}{k-r}\int_{-\infty}^{\infty}\left[\frac{f_i'(y;\boldsymbol{\theta})}{f(y;\boldsymbol{\theta})}\right]^2 f(y;\boldsymbol{\theta})dy<\infty.$$

The proof of (3.17) is trivial and hence omitted. The proof of the lemma is complete.

3.5 Bibliographic notes

The material of this chapter is mainly based on Chen [36] and Bai and Chen [6]. The study on the Fisher information of RSS was first attempted by Stokes [161] for location-scale families. The best linear unbiased estimates in ranked set sampling was also considered by Barnett and Moore [16]. The parametric RSS for particular distributions were considered by many authors. Several examples of the location-scale families and related results on parametric estimation can be found in Fei et al. [55], Lam et al. (1994, 1995), Sinha et al.

[155], and Chuiv and Sinha [47]. Sinha et al. [155] also discussed some variations of the original RSS scheme for normal and exponential distributions. For more related results, we refer the reader to Bhoj [19] [22], Bhoj and Ahsanullah [20], Hossain and Muttlak [61] [62], El-Neweihi and Sinha [53], Lacayo et al. [87], Li and Chuiv [91], Muttlak [103], and Al-Saleh et al. [3]. Finally, we add that there is a limited amount of work on tests of hypotheses based on a ranked set sample. We refer the reader to Shen [147] and Shen and Yuan [148] for testing a normal mean, and to Abu-Dayyeh and Muttlak [1] for testing an exponential mean.

4

Unbalanced Ranked Set Sampling and Optimal Designs

In previous chapters, our discussion has been focused on balanced RSS. In this chapter, we turn to unbalanced RSS. For a given statistical problem, the amount of useful information contained in one order statistic is not necessarily the same as in the others. Thus, it is desirable to quantify the order statistics with the highest amount of information, which gives rise to an unbalanced RSS. In certain practical problems, though RSS does not come into play, the data has the structure of an unbalanced RSS sample. Take, for example, the reliability analysis of the so-called r-out-of-k systems which are of great use for improving reliability by providing component redundancy. An r-out-of-k system consists of k independently functioning identical components. The system functions properly if and only if at least r of its components are still normally functioning. The system fails when the number of its failed components reaches $(k-r+1)$. Thus, the lifetime of an "r-out-of-k" system is, in fact, the $(k-r+1)$st order statistic of k independent identically distributed component lifetimes. Thus, a sample of system lifetimes can be treated as an unbalanced RSS sample while only the $(k-r+1)$st order statistic is quantified in each ranked set of size k, see Kvam and Samaniego [85] and Chen [33]. Another example is the so-called nomination sampling. In certain social surveys, psychological factors often play a role as to whether or not reliable information can be extracted from the sampling units. Nomination sampling, which takes into account the psychological factors, is designed to observe only the maximum or the minimum of a simple random sample. For more details of nomination sampling, the reader is referred to Willemain [170], Tiwari [168] [169].

In this chapter, we deal with the analysis of unbalanced ranked-set data and the design of unbalanced RSS schemes which achieve certain optimalities. A general structure of unbalanced RSS data, which allows different sizes for different ranked sets, is described in Section 4.1. The non-parametric analysis of unbalanced RSS data is discussed in Section 4.2. The non-parametric analysis is based on the sample quantiles of the unbalanced RSS data. A relationship between the distribution function of the population and an weighted average of the distribution functions of the ranked statistics is established, which provides the foundation for making inferences on the population using unbalanced RSS data. The inferences on population quantiles, distribution function and statistical functionals are covered in this section. The parametric analysis of unbalanced RSS data including MLE and best linear unbiased estimate for location-scale families is discussed in Section 4.3. The remaining sections deal with the design of RSS schemes subject to certain optimality criteria. In Section 4.4, the optimal designs for location-scale families using Fisher information matrix are developed. Particular designs with A-optimality and D-optimality are derived for families with Normal and extreme value distributions. In Section 4.5, the optimal designs for the non-parametric estimation of quantiles are developed. The computational issues associated with the optimal RSS schemes are briefly discussed in Section 4.6. The relative efficiency of optimal RSS schemes is illustrated in Section 4.7. Some other design methodologies such as Bayesian design and adaptive design are briefly explored in Section 4.8.

4.1 General structure of unbalanced RSS

An unbalanced RSS is one in which the ranked order statistics are not quantified the same number of times. An unbalanced ranked-set sample is given as follows.

$$\begin{array}{cccc} X_{(1)1} & X_{(1)2} & \cdots & X_{(1)n_1} \\ X_{(2)1} & X_{(2)2} & \cdots & X_{(2)n_2} \\ \cdots & \cdots & \cdots & \cdots \\ X_{(k)1} & X_{(k)2} & \cdots & X_{(k)n_k} \end{array} \qquad (4.1)$$

where all the $X_{(r)i}$'s are independent and the $X_{(r)i}$'s with the same r are also identically distributed. The sample can be considered as obtained from $n = \sum_{r=1}^{k} n_r$ sets of random sampling units, each of size k, by first ranking the units of each set and then, for $r = 1, \cdots, k$, measuring the rth ranked order statistic for n_r ranked sets.

More generally, we can consider an unbalanced ranked-set sample which consists of several copies of sample (4.1), each copy with a different set size. A general unbalanced ranked-set sample can be represented as

$$\begin{array}{cccc} X_{(r_1:k_1)1} & X_{(r_1:k_1)2} & \cdots & X_{(r_1:k_1)n_1} \\ X_{(r_2:k_2)1} & X_{(r_2:k_2)2} & \cdots & X_{(r_2:k_2)n_2} \\ \cdots & \cdots & \cdots & \cdots \\ X_{(r_m:k_m)1} & X_{(r_m:k_m)2} & \cdots & X_{(r_m:k_m)n_m}, \end{array} \qquad (4.2)$$

where $(r_i, k_i) \neq (r_l, k_l)$ if $i \neq l$, and each of the $X_{(r_i:k_i)j}, j = 1, \ldots, n_i$, is the r_ith order statistic of a simple random sample of size k_i from the same population.

To avoid unnecessary complexity of notation, the discussion that follows will be confined to the ranked-set sample of form (4.1). The methodology developed for this form can be easily extended to the ranked-set sample of form (4.2). In the remainder of this section, we introduce some definitions.

The empirical distribution function of unbalanced ranked-set sample (4.1) is defined as

$$\begin{aligned}\hat{F}_{\boldsymbol{q}_n}(x) &= \frac{1}{n}\sum_{r=1}^{k}\sum_{j=1}^{n_r} I\{X_{(r)j} \leq x\} \\ &= \sum_{r=1}^{k} q_{nr}\hat{F}_{(r)}(x),\end{aligned}$$

where $q_{nr} = n_r/n$, $\boldsymbol{q}_n = (q_{n1}, \cdots, q_{nk})^T$ and $\hat{F}_{(r)}(x) = \frac{1}{n_r}\sum_{j=1}^{n_r} I\{X_{(r)j} \leq x\}$.

For $0 < p < 1$, the pth unbalanced ranked-set sample quantile is defined as

$$x_{\boldsymbol{q}_n}(p) = \inf\{x : \hat{F}_{\boldsymbol{q}_n}(x) \geq p\}.$$

Arrange the $X_{(r)j}$s in (4.1) from the smallest to the largest and denote them by

$$Z_{n:1} \leq Z_{n:2} \leq \cdots \leq Z_{n:n}.$$

The $Z_{n:i}, i = 1, \cdots, n$, are referred to as the unbalanced ranked-set order statistics.

4.2 Nonparametric analysis of unbalanced ranked-set data

In this section, we discuss the nonparametric analysis of unbalanced ranked set data. The analysis relies very much on the unbalanced ranked set sample quantiles. We present the asymptotic properties of the unbalanced ranked set sample quantiles in Section 4.2.1. Then, the non-parametric estimation of population quantiles and distribution function is discussed in Section 4.2.2. In Section 4.2.3, confidence intervals and hypothesis testing for quantiles are considered. Section 4.2.4 deals with the estimation of statistical functionals. Section 4.2.5 discusses inference using bootstrap methods. Section 4.2.6 presents some simulation results.

4.2.1 Asymptotic properties of unbalanced ranked-set sample quantiles

As in the previous chapters, let F and f denote the distribution function and density function of the population, $F_{(r)}$ and $f_{(r)}$ denote the distribution function and density function of $X_{(r)}$. The pth quantile of F is denoted by $x(p)$. Suppose that, as $n \to \infty$, $q_{nr} \to q_r$, $r = 1, \ldots, k$. Let $F_{\boldsymbol{q}} = \sum_{r=1}^{k} q_r F_{(r)}$. Note that $F_{\boldsymbol{q}}$ is a distribution function. The pth quantile of $F_{\boldsymbol{q}}$ is denoted by $x_{\boldsymbol{q}}(p)$. The density function of $F_{\boldsymbol{q}}$ is denoted by $f_{\boldsymbol{q}}$.

The asymptotic properties of the unbalanced ranked-set sample quantiles are stated as follows.

Theorem 4.1. *(i) (The strong consistency). With probability 1, $x_{\boldsymbol{q}_n}(p)$ converges to $x_{\boldsymbol{q}}(p)$.*

(ii) (The Bahadur representation). Suppose that $q_{nr} = q_r + O(n^{-1})$. If $f_{\boldsymbol{q}}$ is continuous at $x_{\boldsymbol{q}}(p)$ and positive in a neighborhood of $x_{\boldsymbol{q}}(p)$, then

$$x_{\boldsymbol{q}_n}(p) = x_{\boldsymbol{q}}(p) + \frac{p - \hat{F}_{\boldsymbol{q}_n}(x_{\boldsymbol{q}}(p))}{f_{\boldsymbol{q}}(x_{\boldsymbol{q}}(p))} + R_n,$$

where, with probability one,

$$R_n = O(n^{-3/4}(\log n)^{3/4}),$$

as $n \to \infty$.

(iii)(The asymptotic normality). Under the same assumption as in (ii),

$$\sqrt{n}(x_{\boldsymbol{q}_n}(p) - x_{\boldsymbol{q}}(p)) \to N\left(0, \frac{\sigma^2(\boldsymbol{q}, p)}{f_{\boldsymbol{q}}^2(x_{\boldsymbol{q}}(p))}\right),$$

in distribution, where

$$\sigma^2(\boldsymbol{q}, p) = \sum_{r=1}^{k} q_r F_{(r)}(x_{\boldsymbol{q}}(p))[1 - F_{(r)}(x_{\boldsymbol{q}}(p))].$$

The results above hold whether the ranking in the RSS is perfect or not. The proof of the results can be carried out by the same techniques as in Serfling [146] for simple random sampling and hence is omitted.

4.2.2 Nonparametric estimation of quantiles and distribution function

Assume that the ranking in the RSS is perfect, i.e., the statistic $X_{(r)}$ is indeed the rth order statistic. Under this assumption, we have

$$f_{(r)}(x) = \frac{k!}{(r-1)!(k-r)!} F^{r-1}(x)[1 - F(x)]^{k-r} f(x),$$

and hence

$$F_{(r)}(x) = B(r, k - r + 1, F(x)),$$

where $B(r, s, t)$ denotes the distribution function of the beta distribution with shape parameters r and s. Define

$$s_{\boldsymbol{q}}(t) = \sum_{r=1}^{k} q_r B(r, k - r + 1, t).$$

We then have

$$F_{\boldsymbol{q}}(x) = s_{\boldsymbol{q}}(F(x)). \tag{4.3}$$

In particular, if F is strictly increasing in a neighborhood of $x(p)$, we have

$$F_{\boldsymbol{q}}(x(p)) = s_{\boldsymbol{q}}(F(x(p))) = s_{\boldsymbol{q}}(p).$$

Note that $s_{\boldsymbol{q}}(p)$ is completely determined by p for given $\boldsymbol{q}$. Thus we arrive at the relation $x_{\boldsymbol{q}}(s_{\boldsymbol{q}}(p)) = x(p)$, that is, the p-th quantile of F is the $s_{\boldsymbol{q}}(p)$th quantile of $F_{\boldsymbol{q}}$. Therefore, we can swap the problem of estimating the pth quantile of F for the problem of estimating the $s_{\boldsymbol{q}}(p)$th quantile of $F_{\boldsymbol{q}}$.

The estimate of population quantile. Define the estimate of $x(p)$ as

$$\hat{x}_n(p) = x_{\boldsymbol{q}_n}(s_{\boldsymbol{q}}(p)). \tag{4.4}$$

Note that

$$f_{\boldsymbol{q}}(x(p)) = \sum_{r=1}^{k} q_r \frac{k!}{(r-1)!(k-r)!} p^{r-1}(1-p)^{k-r} f(x(p)).$$

Then (iii) of Theorem 4.1 implies that

$$\sqrt{n}[\hat{x}_n(p) - x(p)] \to N\left(0, \frac{\sigma^2(\boldsymbol{q}, s_{\boldsymbol{q}}(p))}{\tau^2(p) f^2(x(p))}\right),$$

in distribution, where

$$\tau^2(p) = \left[\sum_{r=1}^{k} q_r \frac{k!}{(r-1)!(k-r)!} p^{r-1}(1-p)^{k-r}\right]^2,$$

and

$$\sigma^2(\boldsymbol{q}, s_{\boldsymbol{q}}(p)) = \sum_{r=1}^{k} q_r B(r, k-r+1, p)[1 - B(r, k-r+1, p)].$$

Therefore, the estimate $\hat{x}_n(p)$ of $x(p)$ is asymptotically normally distributed, and, by (i) of Theorem 4.1, it is also strongly consistent.

Now we turn to the estimation of distribution function F. It follows from (4.3) that

$$F(x) = s_{\boldsymbol{q}}^{-1}(F_{\boldsymbol{q}}(x)), \tag{4.5}$$

where $s_{\boldsymbol{q}}^{-1}$ is the inverse function of $s_{\boldsymbol{q}}$. Since $\hat{F}_{\boldsymbol{q}_n}(x)$ is a consistent estimate of $F_{\boldsymbol{q}}(x)$ and $s_{\boldsymbol{q}}^{-1}$ is continuous , an estimate of $F(x)$ can be obtained by substituting $F_{\boldsymbol{q}}$ with $\hat{F}_{\boldsymbol{q}_n}$ in (4.5).

The estimate of distribution function. Define the estimate of $F(x)$ as

$$\hat{F}_n(x) = s_{\boldsymbol{q}}^{-1}(\hat{F}_{\boldsymbol{q}_n}(x)).$$

The estimate $\hat{F}_n(x)$ has the following properties.

Theorem 4.2. *(i) (The strong consistency). Suppose that $F(x)$ is strictly increasing in a neighborhood of x. Then, as an estimate of $F(x)$, $\hat{F}_n(x)$ is strongly consistent.*

(ii) (The asymptotic normality). Under the same assumption of the previous theorem,

$$\frac{\sqrt{n}[\hat{F}_n(x) - F(x)]}{\sigma_F} \to N(0,1),$$

in distribution, where

$$\sigma_F^2 = \frac{\sum_{r=1}^k q_r B(r, k-r+1, F(x))[1 - B(r, k-r+1, F(x))]}{[\sum_{r=1}^k q_r b(r, k-r+1, F(x))]^2}.$$

Here $b(r,s,t)$ denotes the density function of the beta distribution with shape parameters r and s.

(iii) (The uniform convergence). With probability 1,

$$\sup_{-\infty<x<\infty} |\hat{F}_n(x) - F(x)| \to 0.$$

The proof of the theorem is elementary. The details of the proof can be found in Chen [37].

The estimate $\hat{F}_n$ can be explicitly expressed as follows.

$$\hat{F}_n(x) = \sum_{i=1}^n p_i I\{Z_{n:i} \le x\},$$

where $Z_{n:i}$s are the unbalanced ranked set sample order statistics. The weights p_i are determined as follows. Let b_i be determined such that $s_{\boldsymbol{q}}(b_i) = i/n$. Then p_i is determined as $p_1 = b_1$ and $p_i = b_i - b_{i-1}, i = 2, \ldots, n$.

Apart from its own interest, the estimated distribution function has important applications. In a parametric setting, the estimated distribution function can be used to check the validity of the distributional assumption, a task once thought impossible with unbalanced RSS. A histogram can be formed from the estimated distribution function. The usual Q-Q plot can also be carried out. In a nonparametric setting, the so-called plug-in method can be used together with the estimated distribution function for the estimation of other features of the population, as will be seen later.

4.2.3 Confidence interval and hypothesis testing

The procedures for confidence interval and hypothesis testing discussed in Section 2.6.2 can be adapted to the unbalanced RSS.

Suppose that for given p and k, the pth quantile $x(p)$ of F is the sth quantile of $F_{\boldsymbol{q}}$. Then a confidence interval of confidence coefficient $1 - 2\alpha$ for $x(p)$ can be constructed as $[Z_{n:l_1}, \ Z_{n:l_2}]$, where

$$l_1 = ns - z_\alpha[n\sum_{r=1}^{k} q_r B(r, k-r+1, p)[1 - B(r, k-r+1, p)]]^{1/2},$$

$$l_2 = ns + z_\alpha[n\sum_{r=1}^{k} q_r B(r, k-r+1, p)[1 - B(r, k-r+1, p)]]^{1/2},$$

where z_α denotes the $(1-\alpha)$th quantile of the standard normal distribution.

To test $H_0 : x(p) = \xi_0$, the test statistic can be constructed as

$$Z_n = \frac{\sqrt{n}\hat{f}_{\boldsymbol{q}}(\xi_0)[\hat{x}_{\boldsymbol{q}_n}(p) - \xi_0]}{[\sum_{r=1}^{k} q_r B(r, k-r+1, p)[1 - B(r, k-r+1, p)]]^{1/2}},$$

where $\hat{f}_{\boldsymbol{q}}$ is taken as a kernel estimate based on the unbalanced ranked-set sample. The kernel estimate is constructed as follows.

$$\hat{f}_{\boldsymbol{q}}(x) = \sum_{i=1}^{n} p_i \frac{1}{h} K\left(\frac{x - Z_{n:i}}{h}\right), \tag{4.6}$$

where h is the user-determined bandwidth. The estimate (4.6) is a straightforward extension of the kernel estimate discussed in Section 2.6. Under the null hypothesis, the test statistic has approximately a standard normal distribution.

4.2.4 Estimation of statistical functionals*

A statistical functional is of the form $\theta(F)$ which is a function defined at every distribution function F. Trivial examples of statistical functional are mean, variance, etc.. A straightforward method for estimating a statistical functional is the so-called plug-in method, which estimates $\theta(F)$ by $\theta(\hat{F})$ where $\hat{F}$ is an estimate of the distribution function F. With a general unbalanced ranked-set sample, we can take $\hat{F}$ as $\hat{F}_n$ defined Section 4.2.2. We call the plug-in estimator with $\hat{F}_n$ in place the naive estimator for reasons that will become clear later. Note that, like the usual empirical distribution, $\hat{F}_n$ puts all probability mass on the data points, however, the mass at each data point is not necessarily equal. The resultant estimator of $\theta(F)$ will be a weighted version of the usual plug-in estimator. For example, for the population mean $\mu(F) = \int x dF(x)$, the estimator is given by

$$\mu(\hat{F}_n) = \sum_{i=1}^{n} p_i Z_{n:i}.$$

For the population variance $\sigma^2(F) = \int (x - \mu(F))^2 dF(x)$, the estimate is given by

$$\sigma^2(\hat{F}_n) = \sum_{i=1}^{n} p_i [Z_{n:i} - \mu(\hat{F}_n)]^2.$$

In order to get some insights into the nature of the naive estimator, let us consider the properties of p_i in more detail. Suppose that r_1 is the smallest r and r_2 is the largest r such that $q_r > 0$. When i is close to 1, we have $s(b_i) \approx s'(b_i)b_i = b_i^{r_1} + o(b_i^{r_1})$. Thus we have, approximately, $b_i = (i/n)^{1/r_1}$, and hence, $p_i = (1/n)(i/n)^{-(r_1-1)/r_1}$. Similarly, we can derive that, when i is close to n, $p_i = (1/n)(1-i/n)^{-(r_2-1)/r_2}$. However, for medium i, $p_i = (1/n)c_i$ where c_i is bounded. It is then clear that the ratios of the p_i's with i close either to 1 or n and those p_i's with i in the middle of the range go to infinity as n goes to infinity. That is, relatively much larger weights are put on the extreme order statistics, which has an adverse effect on the behavior of the naive estimator and also hinders the derivation of large sample properties. This gives rise to the conception of trimmed plug-in estimator to be defined below. Let $0 < \alpha, \beta < 1/2$. Let

$$\hat{F}_{n,\alpha,\beta}(x) = \frac{\hat{F}_n(x) - \hat{F}_n(\hat{x}_n(\alpha))}{1-\alpha-\beta} I\{\hat{x}_n(\alpha) \le x \le \hat{x}_n(1-\beta)\}.$$

Then we define the trimmed plug-in estimator of $\theta(F)$ as $\theta(\hat{F}_{n,\alpha,\beta})$.

In what follows, we investigate the asymptotic properties of the trimmed plug-in estimator. For convenience, we will confine to the estimator of population mean. First, we define a trimmed version of the population distribution similarly as

$$F_{\alpha,\beta}(x) = \frac{F(x) - F(x(\alpha))}{1-\alpha-\beta} I\{x(\alpha) \le x \le x(1-\beta)\}.$$

For the mean functional, we have

$$\begin{aligned}
\theta(F_{\alpha,\beta}) &= \int_{-\infty}^{\infty} \frac{xI\{x(\alpha) \le x \le x(1-\beta)\}}{1-\alpha-\beta} dF(x) \\
&= \int_{-\infty}^{\infty} \frac{xI\{x(\alpha) \le x \le x(1-\beta)\}}{1-\alpha-\beta} ds^{-1}(F_{\boldsymbol{q}}(x)) \\
&= \int_0^1 F_{\boldsymbol{q}}^{-1}(t) \frac{s^{-1'}(t) I\{s(\alpha) \le t \le s(1-\beta)\}}{1-\alpha-\beta} dt \\
&= \int_0^1 F_{\boldsymbol{q}}^{-1}(t) J(t) dt,
\end{aligned}$$

where

$$J(t) = \frac{s^{-1'}(t) I\{s(\alpha) \le t \le s(1-\beta)\}}{1-\alpha-\beta},$$

and s is the abbreviation for $s_{\boldsymbol{q}}$. Here both $F_{\boldsymbol{q}}^{-1}$ and s^{-1} denote inverse functions rather than reciprocals. Thus, we can consider $\theta(F_{\alpha,\beta})$ as a functional of $F_{\boldsymbol{q}}$. Let the functional be denoted by $T(F_{\boldsymbol{q}})$. It is easy to derive that the influence curve of the functional $T(\cdot)$ at $F_{\boldsymbol{q}}$ is given by

$$IC(x;T,F_{\boldsymbol{q}}) = -\int_{-\infty}^{\infty}[I\{x \le y\} - F_{\boldsymbol{q}}(y)]J(F_{\boldsymbol{q}}(y))dy.$$

Note that as long as the asymptotic properties are of concern we can replace the empirical quantiles $\hat{x}_n(\alpha)$ and $\hat{x}_n(1-\beta)$ by the quantiles of the population in the definition of $\hat{F}_{n,\alpha,\beta}$ since the empirical quantiles converge to the population quantiles at a rate of $O(n^{-1/2})$. Therefore, we have

$$\theta(\hat{F}_{n,\alpha,\beta}) = T(\hat{F}_{\boldsymbol{q}_n}).$$

It follows that

$$\begin{aligned}
&\sqrt{n}[\theta(\hat{F}_{n,\alpha,\beta}) - \theta(F_{\alpha,\beta})]\\
&= \sqrt{n}[T(\hat{F}_{\boldsymbol{q}_n}) - T(F_{\boldsymbol{q}})]\\
&= -\frac{1}{\sqrt{N}}\sum_{r=1}^{k}\sum_{j=1}^{n_r}\int_{-\infty}^{\infty}[I\{X_{(r)j} \le y\} - F_{\boldsymbol{q}}(y)]J(F_{\boldsymbol{q}}(y))dy + \sqrt{n}\Delta_n\\
&= -\sum_{r=1}^{k}\sqrt{q_r}\frac{1}{\sqrt{n_r}}\sum_{j=1}^{n_r}\int_{-\infty}^{\infty}[I\{X_{(r)j} \le y\} - F_{(r)}(y)]J(F_{\boldsymbol{q}}(y))dy + \sqrt{n}\Delta_n.
\end{aligned}$$

It can be proved that $\sqrt{n}\Delta_n \to 0$ in probability, see Chen [37]. Hence, by the central limit theorem, we have the following result:

Theorem 4.3. *(The asymptotic normality of trimmed plug-in estimator) Suppose that $0 < \alpha, \beta < 1/2$ and that F is strictly increasing at both $x(\alpha)$ and $x(1-\beta)$. Then*

$$\sqrt{n}[\theta(\hat{F}_{n,\alpha,\beta}) - \theta(F_{\alpha,\beta})] \to N(0, \sigma^2(\boldsymbol{q})),$$

in distribution, where

$$\sigma^2(\boldsymbol{q}) = \sum_{r=1}^{k} q_r \sigma^2(T, F_{(r)})$$

with

$$\sigma^2(T, F_{(r)}) = Var\left(\int_{-\infty}^{\infty}[I\{X_{(r)} \le y\} - F_{(r)}(y)]J(F_{\boldsymbol{q}}(y))dy\right).$$

Although the asymptotic normality is derived only for the trimmed plug-in estimator of mean, it holds for a large class of estimators. Specifically, let $\theta(F) = g(m_1(F), \ldots, m_k(F))$, where g is a smooth function and $m_j(F)$ is the jth moment of F. Then $\theta(\hat{F}_{n,\alpha,\beta})$, as an estimator of $\theta(F)$, will be asymptotically normally distributed. However, it must be noticed that the trimmed plug-in estimator is not necessarily asymptotically unbiased in general. The bias is affected by the choice of α and β. In the special case of estimating the mean, if F is a symmetric distribution, the α and β can be chosen equal to make the trimmed plug-in estimator asymptotically unbiased. However, in general cases, bias correction has to be made through some other means if necessary.

4.2.5 Inference by bootstrapping

In the previous subsection, we derived the asymptotic distribution of the trimmed plug-in estimators. However, the asymptotic distribution of the trimmed plug-in estimator cannot be readily used for this inference since there is no appropriate estimate of the asymptotic variance. We have to resort to other means such as bootstrapping for this purpose.

The bootstrap method proposed by Efron [52] consists of two layers of procedures: to estimate a quantity $G(F)$ by the plug-in estimator $G(\hat{F})$ and to compute $G(\hat{F})$ by simulation if necessary. For example, $G(F)$ could be the bias of a statistic T_n, an estimator of a parameter $\theta(F)$. In this case, $G(F) = E(T_n|F) - \theta(F)$. $G(F)$ could also be the distribution of a certain random quantity, say, $T_n - \theta(F)$. Then $G(F) = P(T_n - \theta(F) \leq x|F)$. In both cases, $G(\hat{F})$ cannot be computed directly and a Monte-Carlo simulation is in order. Since the unbalanced ranked-set sample (4.1) can be considered as k independent simple random samples from the distributions $F_{(r)}$, a bootstrap sample (or a re-sample) can be generated by separately resampling from the k simple random samples. This gives rise to the following bootstrap procedure for the estimation of the distribution of $T_n = \sqrt{n}[\theta(\hat{F}_{n,\alpha,\beta}) - \theta(F_{\alpha,\beta})]$.

For $l = 1, \ldots, B$, do the following:

- *For $r = 1, \ldots, k$, sample with replacement n_r times from $\{X_{(r)1}, \cdots, X_{(r)n_r}\}$ to obtain the resample : $\{X^*_{(r)1}, \cdots, X^*_{(r)n_r}\}$.*
- *From the bootstrap sample $\{X^*_{(r)j}, r = 1, \cdots, k; j = 1, \cdots, n_r\}$, form, in turn, $\hat{\hat{F}}_{\mathbf{q}_n}$, $\hat{\hat{F}}_n$ and $\hat{\hat{F}}_{n,\alpha,\beta}$, and compute*

$$T^*_{nl} = \sqrt{n}[\theta(\hat{\hat{F}}_{n,\alpha,\beta}) - \theta(\hat{F}_{n,\alpha,\beta})].$$

Finally, form the empirical distribution of $T^*_{nl}, l = 1, \ldots, B$.

The empirical distribution of T^*_{nl}'s is then used as the approximation to the distribution of $\sqrt{n}[\theta(\hat{\hat{F}}_{n,\alpha,\beta}) - \theta(\hat{F}_{n,\alpha,\beta})]$ which, in turn, is used as the estimate of the distribution of $\sqrt{n}[\theta(\hat{F}_{n,\alpha,\beta}) - \theta(F_{\alpha,\beta})]$.

4.2.6 Simulation results

In order to illustrate the results of the previous subsections some simulation studies are presented in the following. The estimation of the mean and the distribution function is simulated for the Normal, Weibull and Extreme Value distributions. The simulation is done as follows. For a given distribution and a given data configuration, 1000 general unbalanced ranked-set samples are generated by using S-plus 4.5. For each of the samples, the estimates of mean and distribution function and the average squared error (ASE) of the

Table 4.1. Data configurations

1		2		3		4	
$r:k$	Nq_r	$r:k$	Nq_r	$r:k$	Nq_r	$r:k$	Nq_r
1:10	12	1:10	10	1:5	10	(a)	
2:10	5	2:10	9	3:5	9	3:5	25
3:10	7	3:10	10	5:5	11	4:7	20
4:10	9	4:10	11	1:7	12	(b)	
5:10	10	5:10	7	4:7	10	3:5	30
6:10	13	6:10	11	7:7	13	4:7	35
7:10	14	7:10	11			(c)	
8:10	11	8:10	14			3:5	45
9:10	11	9:10	7			4:7	45
10:10	11	10:10	10				

Table 4.2. Summary of simulation results

Distri-bution	Data Config.	Popu. mean	AVG.m	MSE.m	IMSE.df
Normal	1	0.5	0.498	6.343e-06	0.00065
Weibull	2	0.4431	0.4428	0.00012	0.00067
Weibull	3	0.4431	0.4467	0.00029	0.00131
Extreme V.	4(a)	-0.5772	-0.5429	0.01197	0.00161
Extreme V.	4(b)	-0.5772	-0.5556	0.00897	0.00110
Extreme V.	4(c)	-0.5772	-0.5616	0.00718	0.00080

distribution function estimate are computed. Then the estimated expectation and the mean square error of the mean estimate are computed respectively as AVG.m $= (1/1000)\sum_{j=1}^{1000} \hat{\mu}_j$ and MSE.m $= (1/1000)\sum_{j=1}^{1000}(\hat{\mu}_j - \mu)^2$ where μ is the true mean of the underlying distribution. The integrated mean square error of the distribution function estimates is computed as IMSE.df $= (1/1000)\sum_{j=1}^{1000} ASE_j$ where for each sample ASE is computed as $(1/n)\sum_{i=1}^{n}(\hat{F}(Z_{n:i}) - F(Z_{n:i}))^2$.

The following four settings are taken in the simulation: (i) Normal distribution with mean 0.5 and variance 0.04 and data configuration 1, (ii) Weibull distribution with shape parameter 2 and scale parameter 0.5 and data configuration 2, (iii) the same Weibull distribution as in (ii) but with data configuration 3, (iv) standard Extreme Value distribution with data configurations 4 (a)—(c). The data configurations are given in Table 4.1. The first two configurations have a fixed set size $k = 10$. Configuration 3 represents simulated r-out-of-k systems including series structures and parallel structures as special cases. Configurations 4(a)—4(c) represent optimal ranked-set sampling schemes for the estimation of population median, c.f., Section 4.5. The simulation results are summarized in Table 4.2. These results illustrate that the non-parametric approach provides very good estimates for the mean and distribution function even if the data configuration is very unbalanced as in configuration 4.

4.3 Parametric analysis of unbalanced ranked-set data

In this section, we discuss the analysis of unbalanced ranked-set data when the underlying distribution is known up to certain parameters. We consider two methods: maximum likelihood estimation (MLE) and best linear unbiased estimation (BLUE).

Maximum likelihood estimation. Let $f(x, \boldsymbol{\theta})$ and $F(x, \boldsymbol{\theta})$ be the density function and distribution function of the underlying distribution, where $\boldsymbol{\theta}$ is the vector of the unknown parameters. The density function of the rth order statistic is then given by

$$f_{(r)}(x, \boldsymbol{\theta}) = \frac{k!}{(r-1)!(k-r)!} F^{r-1}(x, \boldsymbol{\theta})[1 - F(x, \boldsymbol{\theta})]^{k-r} f(x, \boldsymbol{\theta}).$$

Given the unbalanced ranked-set data (4.1), the log likelihood function of $\boldsymbol{\theta}$ is obtained as

$$l(\boldsymbol{\theta}) = \sum_{r=1}^{k} \sum_{i=1}^{n_r} \log(f_{(r)}(X_{(r)i}, \boldsymbol{\theta})).$$

In principle, the MLE of $\boldsymbol{\theta}$ based on the unbalanced ranked-set data can be obtained by maximizing $l(\boldsymbol{\theta})$. As in the case of balanced RSS, the MLE is strongly consistent and asymptotically normally distributed.

Let $I_r(\boldsymbol{\theta})$ denote the information matrix of a single $X_{(r)j}$. We have

$$I_r(\boldsymbol{\theta}) = E\left[\frac{\partial \ln f_{(r)}}{\partial \boldsymbol{\theta}} \frac{\partial \ln f_{(r)}}{\partial \boldsymbol{\theta}^T}\right],$$

where

$$\frac{\partial \ln f_{(r)}}{\partial \boldsymbol{\theta}} = \left(\frac{\partial \ln f_{(r)}}{\partial \theta_1}, \dots, \frac{\partial \ln f_{(r)}}{\partial \theta_q}\right)^T,$$

and $\frac{\partial \ln f_{(r)}}{\partial \boldsymbol{\theta}^T}$ is the transpose of $\frac{\partial \ln f_{(r)}}{\partial \boldsymbol{\theta}}$. The information matrix based on data (4.1) is then given by

$$I(\boldsymbol{\theta}; \boldsymbol{n}) = \sum_{r=1}^{k} n_r I_r(\boldsymbol{\theta}) \tag{4.7}$$

where $\boldsymbol{n} = (n_1, \dots, n_k)'$. Suppose that $n_r/n \to q_r, r = 1, \cdots, k$. Let $\hat{\boldsymbol{\theta}}_{ML}$ denote the MLE of $\boldsymbol{\theta}$ based on data (4.1). Then, under the regularity conditions,

$$\sqrt{n}(\hat{\boldsymbol{\theta}}_{ML} - \boldsymbol{\theta}) \to N(0, I(\boldsymbol{\theta}, \boldsymbol{q})^{-1}),$$

in distribution, where

$$q = (q_1, \ldots, q_k)'$$
$$I(\theta; q) = \sum_{r=1}^{k} q_r I_r(\theta). \tag{4.8}$$

This result can be used to construct confidence intervals or conduct hypotheses testing about θ.

Best linear unbiased estimation. As in the balanced RSS case, if the underlying distribution belongs to a location-scale family, the best linear unbiased estimates can be derived for the location and scale parameters of the underlying distribution.

Recall that we can express the $X_{(r)i}$'s in (4.1) as follows.

$$X_{(r)i} = \mu + \sigma\alpha_r + \epsilon_{ri}, \qquad r = 1, \ldots, k; i = 1, \ldots, n_r, \tag{4.9}$$

where ϵ_{ri} are independent random variables with $E\epsilon_{ri} = 0$ and $\mathrm{Var}(\epsilon_{ri}) = \sigma^2\nu_r$. Here $\alpha_r = E\frac{X_{(r)i}-\mu}{\sigma}$ and $\nu_r = \mathrm{Var}(\frac{X_{(r)i}-\mu}{\sigma})$. The BLUEs of μ and σ are given by their weighted least squares estimates with weights $w_{ri} = \nu_r^{-1}$ based on (4.9). Note that the unbiasedness of the estimates does not depend on whether the RSS is balanced or not.

The weighted least squares estimates of σ and μ are given, respectively, by

$$\tilde{\sigma}_{\text{BLUE}} = \frac{\sum_{r=1}^{k} \frac{q_{nr}}{\nu_r}(\alpha_r - \tilde{\alpha})\bar{X}_{(r)}}{\sum_{r=1}^{k} \frac{q_{nr}}{\nu_r}(\alpha_r - \tilde{\alpha})^2},$$
$$\tilde{\mu}_{\text{BLUE}} = \tilde{X}_W - \tilde{\alpha}\tilde{\sigma},$$

where

$$\tilde{\alpha} = \sum_{r=1}^{k} q_{nr}\alpha_r, \; \bar{X}_{(r)} = \frac{1}{n_r}\sum_{i=1}^{n_r} X_{(r)i}, \; \tilde{X}_W = \frac{\sum_{r=1}^{k} \frac{q_{nr}}{\nu_r}\bar{X}_{(r)}}{\sum_{r=1}^{k} \frac{q_{nr}}{\nu_r}}.$$

If, when $n = \sum_{r=1}^{k} n_r \to \infty$, $q_{nr} \to q_r, r = 1, \cdots, k$, then we have that the joint distribution of $\tilde{\mu}_{\text{BLUE}}$ and $\tilde{\sigma}_{\text{BLUE}}$ converges to a bivariate normal distribution with mean (μ, σ) and an asymptotic variance-covariance matrix given by $\sigma^2 V^{-1}(q)$ where

$$V(q) = \begin{pmatrix} \sum_{r=1}^{k} \frac{q_r}{\nu_r} & \sum_{r=1}^{k} \frac{q_r\alpha_r}{\nu_r} \\ \sum_{r=1}^{k} \frac{q_r\alpha_r}{\nu_r} & \sum_{r=1}^{k} \frac{q_r\alpha_r^2}{\nu_r} \end{pmatrix}. \tag{4.10}$$

In certain practical problems, the data just come out naturally to have an unbalanced RSS structure, as in the cases of r-out-of-k systems and nomination sampling mentioned in the beginning of this chapter, and SRS is not a practical alternative. Therefore a comparison of such a given unbalanced RSS structure with the SRS is practically meaningless. However, we point out that the estimates based on a given unbalanced RSS is not necessarily more efficient than their counterparts based on a balanced RSS or on an SRS. Whenever we are at the discretion on how to sample, we should compare the efficiency of all possible sampling schemes and find the one which is the most efficient. This naturally leads to the problem of the design of optimal RSS schemes which we take up in the rest of this chapter.

4.4 Optimal design for location-scale families

An optimal design of RSS is an allocation of quantifications among the k ranked order statistics so that the resultant ranked set sample can achieve certain optimality. We take the asymptotic variance and the like as the optimality criteria for the estimation of parameters. The optimal design for location-scale families is considered in this section. A location-scale family is a family of distributions with distribution function of the form $F(\frac{x-\mu}{\sigma})$ where F is a completely known distribution function and μ and σ are, respectively, location and scale parameters. The optimal designs for the estimation of μ and σ are dealt with in this section. Section 4.4.1 concerns optimal designs for maximum likelihood estimation (MLE). Section 4.4.2 discusses optimal designs for best linear unbiased estimation (BLUE).

4.4.1 Optimal RSS for MLE

First, we develop a general methodology of optimal designs for parametric families. Consider parametric families with distribution function $F(x;\boldsymbol{\theta})$ and density function $f(x;\boldsymbol{\theta})$, where $\boldsymbol{\theta} = (\theta_1, \ldots, \theta_q)'$ is a vector of unknown parameters, and assume that the regularity conditions on $f(x;\boldsymbol{\theta})$ as given in the appendix of Chapter 3 are satisfied. We also assume that the ranking in RSS is perfect.

Recall that, given an unbalanced RSS which allocates n_r quantifications to the rth ranked order statistics for $r = 1, \cdots, k$, the Fisher information matrix on $\boldsymbol{\theta}$ based on the resultant data is $I(\boldsymbol{\theta}, \boldsymbol{n})$ as given by (4.7). Under the assumption that $n_r/n \to q_r, r = 1, \cdots, k$, $I(\boldsymbol{n}, \boldsymbol{\theta})/n$ converges to $I(\boldsymbol{\theta}, \boldsymbol{q})$ as given by (4.8). The asymptotic variance-covariance matrix of the MLE of $\boldsymbol{\theta}$, which is the inverse of $I(\boldsymbol{\theta}, \boldsymbol{q})$, is then a function of the allocation proportions q_r for fixed $\boldsymbol{\theta}$. Optimality criteria can be defined as certain functions of the asymptotic variance-covariance matrix. When the simultaneous estimation of the components of $\boldsymbol{\theta}$ is of concern, we consider the following two criteria: the determinant and the trace of the asymptotic variance-covariance matrix denoted by, respectively, $|I(\boldsymbol{\theta}, \boldsymbol{q})^{-1}|$ and $tr(I(\boldsymbol{\theta}, \boldsymbol{q})^{-1})$. A scheme that minimizes $|I(\boldsymbol{\theta}, \boldsymbol{q})^{-1}|$ with respect to q_r's is referred to as the D-optimal scheme. The D-optimal scheme renders the area of the confidence region for $\boldsymbol{\theta}$ based on the asymptotic distribution of $\hat{\boldsymbol{\theta}}_{ML}$ the smallest. A scheme that minimizes $tr(I(\boldsymbol{\theta}, \boldsymbol{q})^{-1})$ with respect to q_r's is referred to as the A-optimal scheme. The A-optimal scheme, in fact, minimizes the sum of the asymptotic variances of the components of $\hat{\boldsymbol{\theta}}_{ML}$. If the estimation of some special function of $\boldsymbol{\theta}$, say $g(\boldsymbol{\theta})$, is of concern, we consider the scheme that minimizes the asymptotic variance of $g(\hat{\boldsymbol{\theta}})$ which is given by $\frac{\partial g}{\partial \boldsymbol{\theta}'} I(\boldsymbol{\theta}, \boldsymbol{q})^{-1} \frac{\partial g}{\partial \boldsymbol{\theta}}$. In particular, if $g(\boldsymbol{\theta}) = \theta_i$, the i-th component of $\boldsymbol{\theta}$, the optimal scheme minimizes the i-th diagonal element of $I(\boldsymbol{\theta}, \boldsymbol{q})^{-1}$.

In general, since the optimality criteria considered above depend on $\boldsymbol{\theta}$, the criteria are not practical without certain knowledge about $\boldsymbol{\theta}$. However, in the special case of location-scale families, the matter becomes much simpler. In fact, the optimization of the criteria does not involve the unknown parameters at all.

Let $f(y) = F'(y)$. Let

$$f_{(r)}(y) = \frac{k!}{(r-1)!(k-r)!} F^{r-1}(y)[1 - F(y)]^{k-r} f(y).$$

Then the density function of the r-th order statistic of a sample of size k from $F(\frac{x-\mu}{\sigma})$ can be written as

$$\frac{1}{\sigma} f_{(r)} \left(\frac{x - \mu}{\sigma} \right).$$

It follows after some straightforward calculation that the information matrix of the r-th order statistic is given by

$$I_r(\mu, \sigma) = \frac{1}{\sigma^2} \begin{pmatrix} A_{(r)11} & A_{(r)12} \\ A_{(r)21} & A_{(r)22} \end{pmatrix}, \tag{4.11}$$

where

$$\begin{aligned} A_{(r)11} &= E \left[\frac{f'_{(r)}(X)}{f_{(r)}(X)} \right]^2, \\ A_{(r)22} &= E \left[\frac{f'_{(r)}(X)}{f_{(r)}(X)} X + 1 \right]^2, \\ A_{(r)12} &= A_{(r)21} \\ &= E \left[\frac{f'_{(r)}(X)}{f_{(r)}(X)} \right] \left[\frac{f'_{(r)}(X)}{f_{(r)}(X)} X + 1 \right]. \end{aligned}$$

The expectations here are taken with respect to the density function $f_{(r)}(y)$ that is free of μ and σ.

Denote the matrix in the right hand side of (4.11) by $A_{(r)}$. Let

Table 4.3. The optimal RSS schemes for Normal distributions

set	*D*-optimality		*A*-optimality	
size	Orders	proportions	Orders	proportions
3	1	0.5	1, 2	0.45, 0.1
4	1	0.5	1, 2	0.38, 0.12
5	1, 2	0.49, 0.01	1, 2	0.31, 0.19
6	1, 2	0.37, 0.13	1, 2	0.23, 0.27
7	1, 2	0.24, 0.26	1, 2	0.13, 0.37
8	1, 2	0.1, 0.4	1, 2	0.03, 0.47
9	2	0.5	2	0.5
10	2	0.5	2	0.5

$$A(\boldsymbol{q}) = \sum_{r=1}^{k} q_r A_{(r)}.$$

Thus, we have

$$I((\mu, \sigma), \boldsymbol{q}) = \frac{1}{\sigma^2} A(\boldsymbol{q}).$$

Then D-optimality and A-optimality are equivalent to, respectively, the minimization of the determinant and the trace of $[A(\boldsymbol{q})]^{-1}$. After the optimal $\boldsymbol{q}$ is found, the n_r is determined as $n_r = [nq_r]$, where $[x]$ denotes the integer closest to x.

In the following, we present the optimal schemes for some important distribution families.

The Normal distributions. In the case that only the location parameter μ is of interest, the optimal RSS scheme minimizes the asymptotic variance of the MLE of μ. The optimal scheme prescribes to quantify the median. If k is odd, the order statistic $X_{(\frac{k+1}{2})}$ is quantified for all sets. If k is even, the order statistic $X_{(\frac{k}{2})}$ is quantified for half of the sets and the order statistic $X_{(\frac{k}{2}+1)}$ is quantified for the other half of the sets.

In the case that only the scale parameter σ is of interest, the optimal RSS scheme minimizes the asymptotic variance of the MLE of σ. The optimal scheme prescribes to quantify the smallest order statistic $X_{(1)}$ for half of the sets and the largest order statistic $X_{(k)}$ for another half of the sets.

For the simultaneous estimation of the location and scale parameters, the D-optimal and the A-optimal schemes with $k = 3, \ldots, 10$, are presented in Table 4.3. There is a symmetric structure of the optimal schemes, that is, if an optimal scheme puts a proportion q on order statistic $X_{(r)}$, it also puts the same proportion on order statistic $X_{(k-r+1)}$. In the table, only the proportions for the lower half of the order statistics to be quantified are presented.

The D-optimality and the A-optimality criteria do not result in the same sampling scheme in general. But both optimal schemes quantify order statistics at the tails. At each tail one or two consecutive order statistics are

Table 4.4. The optimal RSS schemes for estimating only one of the parameters of extreme value distributions

set	Location		Scale	
size	Orders	Proportions	Orders	Proportions
3	2, 3	0.81, 0.19	1, 3	0.57, 0.43
4	3	1	1, 4	0.61, 0.39
5	3, 4	0.46, 0.54	1, 5	0.64, 0.36
6	4, 5	0.9, 0.1	1, 6	0.66, 0.34
7	4, 5	0.15, 0.85	1, 7	0.67, 0.33
8	5, 6	0.59, 0.41	1, 8	0.68, 0.32
9	6	1	2, 9	0.62, 0.38
10	6, 7	0.29, 0.71	2, 10	0.63, 0.37

quantified. As the set size k gets larger, the ranks of the quantified order statistics shrink slowly towards the center. This can be demonstrated by the computation for larger set sizes.

The Extreme value distributions. The cumulative distribution function of an extreme value distribution is given by

$$F\left(\frac{x-\mu}{\sigma}\right) = 1 - \exp\left[-\exp\left(\frac{x-\mu}{\sigma}\right)\right].$$

The extreme value distribution can be obtained as the distribution of the log of a Weibull distributed random variable. It plays an important role in reliability analysis. The extreme value distribution is not symmetric. The optimal RSS schemes for estimating one parameter only lack the simplicity enjoyed by those for normal distributions. The optimal RSS schemes with $k = 3, \ldots, 10$ in the estimation of the location parameter μ only and the scale parameter σ only are presented in Table 4.4.

For the estimation of the location parameter of the extreme value distribution, the weighted average of the orders in Table 4.4, weighted by their respective quantification proportions, can be considered as an effective equivalent order. For $k = 3(1)10$, the effective equivalent orders are, respectively, 2.19, 3, 3.54, 4.1, 4.85, 5.41, 6 and 6.71. When these effective orders are increased by 0.5 and then divided by $k+1$, a common correction used for the estimation of probabilities, we obtain, after rounding to two digits, the following numbers: 0.67, 0.70, 0.67, 0.66, 0.67, 0.66, 0.65 and 0.66. It is interesting to compare these numbers with the mass, 0.632, of the standard extreme value distribution to the left of μ (zero). Then it is clear that the optimal RSS schemes tend to quantify the order statistics near μ, the point of interest, which is in line with the situation of Normal distributions. In the estimation of the scale parameter, the optimal RSS schemes still quantify the order statistics at two tails but with higher proportion at the tail that is heavier.

Table 4.5. The optimal RSS schemes for extreme value distributions

set	*D*-optimality		*A*-optimality	
size	Orders	Proportions	Orders	Proportions
3	1, 3	0.156, 0.84	1, 2, 3	0.12, 0.49, 0.39
4	2, 4	0.33, 0.67	2, 4	0.58, 0.42
5	2, 5	0.38, 0.62	2, 5	0.56, 0.44
6	2, 6	0.4, 0.6	2, 3, 6	0.4, 0.18, 0.42
7	2, 3, 7	0.22, 0.21, 0.57	2, 3, 7	0.07, 0.52, 0.41
8	3, 8	0.45, 0.55	3, 8	0.58, 0.42
9	3, 9	0.45, 0.55	3, 4, 9	0.48, 0.1, 0.42
10	3, 10	0.45, 0.55	3, 4, 10	0.15, 0.44, 0.41

Table 4.6. The optimal RSS schemes for the BLUE of normal distributions

set	*D*-optimality		*A*-optimality	
size	Orders	proportions	Orders	proportions
3	1,3	0.5,0.5	1,3	0.5,0.5
4	1,4	0.5,0.5	1,4	0.5,0.5
5	1,5	0.5,0.5	1,5	0.5,0.5
6	1,6	0.5,0.5	1,2,6	0.41,0.1,0.49
7	1,2,7	0.49,0.01,0.5	1,2,7	0.23,0.29,0.48
8	1,2,8	0.27,0.24,0.49	1,2,8	0.04,0.48,0.48
9	1,2,9	0.03,0.47,0.5	1,2,8	0.34,0.16,0.5
10	1,2,9	0.27,0.24,0.49	1,2,9	0.13,0.37,0.5

The *D*-optimal and the *A*-optimal RSS schemes with $k = 3, \ldots, 10$, for the estimation of both parameters are presented in Table 4.5. Again, the *D*-optimality and the *A*-optimality criteria do not result in the same RSS scheme in general. The two optimal schemes quantify order statistics on two tails of the distribution but with a difference in proportions of quantification on the tails. The *D*-optimal schemes put more proportion on the lighter tail while the *A*-optimal schemes put more proportion on the heavier tail.

Some general patterns of the optimal schemes for the Normal distribution and the extreme value distribution can now be discussed. For both families, to estimate the location parameter alone, the optimal schemes quantify the order statistics near the location parameter, while to estimate the scale parameter alone, the optimal schemes quantify the order statistics at the tails of the distribution. When location and scale parameters are simultaneously estimated, a compromise must be struck between the two tendencies in the separate estimation of the location and scale. Although we can see the general patterns above, there is no promise of a general neat theoretical formula for the optimal allocation, and a solution for each k has to be found numerically.

Table 4.7. The optimal RSS schemes for the BLUE of extreme value distributions

set	*D*-optimality		*A*-optimality	
size	Orders	proportions	Orders	proportions
3	1,3	0.5,0.5	1,3	0.51,0.49
4	1,4	0.5,0.5	1,4	0.53,0.47
5	1,2,5	0.44,0.06,0.5	1,2,5	0.36,0.2,0.44
6	2,6	0.5,0.5	2,6	0.56,0.44
7	2,7	0.5,0.5	2,7	0.56,0.44
8	2,8	0.5,0.5	2,3,8	0.38,0.19,0.43
9	2,3,9	0.17,0.33,0.5	2,3,9	0.01,0.56,0.43
10	3,10	05,0.5	3,10	0.57,0.43

4.4.2 Optimal RSS for BLUE

In Section 4.3, we derived, for a location-scale family, the BLUEs of the location and scale parameters with a given unbalanced RSS scheme and their asymptotic variance-covariance matrix. Recall that the asymptotic variance-covariance matrix is given by $\sigma^2 V^{-1}(\boldsymbol{q})$ where

$$V(\boldsymbol{q}) = \begin{pmatrix} \sum_{r=1}^{k} \frac{q_r}{\nu_r} & \sum_{r=1}^{k} \frac{q_r \alpha_r}{\nu_r} \\ \sum_{r=1}^{k} \frac{q_r \alpha_r}{\nu_r} & \sum_{r=1}^{k} \frac{q_r \alpha_r^2}{\nu_r} \end{pmatrix}.$$

The matrix $V^{-1}(\boldsymbol{q}))$ does not depend on the unknown parameters. As in the case of MLE, optimal schemes can be determined by minimizing certain functions of $V^{-1}(\boldsymbol{q}))$ with respect to $\boldsymbol{q}$. As an illustration, the *D*-optimal and *A*-optimal schemes for the simultaneous estimation of the location and scale parameters of the Normal distribution and the Extreme value distribution by the BLUE are presented, respectively, in Table 4.6 and Table 4.7.

4.5 Optimal design for estimation of quantiles

In this section, we consider the design of optimal RSS schemes for the non-parametric estimation of population quantiles. The optimal designs for the estimation of a single quantile are discussed in Section 4.5.1 and those for the simultaneous estimation of more than one quantile are discussed in Section 4.5.2.

4.5.1 Optimal RSS scheme for the estimation of a single quantile

As discussed in Section 4.2, given an unbalanced ranked set sample as represented in (4.1), the pth quantile of F can be estimated by the $s_{\boldsymbol{q}}(p)$th quantile of the empirical distribution $\hat{F}_{\boldsymbol{q}_n}$. Theorem 4.1 gives the asymptotic variance of the estimate. These results provide readily a means for the design of optimal RSS schemes for the estimation of a single quantile. The design problem can be formulated as finding the unbalanced RSS such that the asymptotic variance of the corresponding estimate of the quantile attains a minimum. Therefore the design problem amounts to minimizing the asymptotic variance with respect to $\boldsymbol{q}$. Recall that the asymptotic variance of the estimate with allocation vector $\boldsymbol{q}$ is given by $V(\boldsymbol{q},p)/f^2(\xi_p)$ where

$$V(\boldsymbol{q},p) = \frac{\sum_{r=1}^k q_r B(r,k-r+1,p)[1-B(r,k-r+1,p)]}{[\sum_{r=1}^k q_r b(r,k-r+1,p)]^2}. \tag{4.12}$$

See Section 4.2.2. Hence, we can find, by minimizing $V(\boldsymbol{q},p)$ with respect to $\boldsymbol{q}$, an unbalanced RSS scheme which results in an asymptotically unbiased minimum variance estimator among all ranked-set sample quantiles. The algorithm for determining the optimal allocation and the corresponding $s_{\boldsymbol{q}}(p)$ is as follows.

Step 1. Determination of the optimal allocation: Minimize $V(\boldsymbol{q},p)$ with respect to $\boldsymbol{q}$ and derive the minimizer $\boldsymbol{q}^* = (q_1^*,\ldots,q_k^*)'$. The allocation is determined as $n_r = [nq_r^*], r = 1,\ldots,k$.

Step 2. Determination of $s_{\boldsymbol{q}^*}(p)$: $s_{\boldsymbol{q}^*}(p) = F_{\boldsymbol{q}^*}(\xi_p) = \sum q_r^* B(r,k-r+1,p)$.

In the following, we consider some properties of $V(\boldsymbol{q},p)$ and $s_{\boldsymbol{q}}(p)$ that give rise to some desirable properties of the optimal schemes.

Lemma 4.4. *Let $\boldsymbol{q}$ be any allocation vector and $\tilde{\boldsymbol{q}}$ be the vector whose rth element $\tilde{q}_r$ equals the $(k-r+1)$st element, q_{k-r+1}, of $\boldsymbol{q}$. Then we have, for $p=0.5$, $V(\boldsymbol{q},p) = V(\tilde{\boldsymbol{q}},p) = V(\frac{1}{2}(\boldsymbol{q}+\tilde{\boldsymbol{q}}),p)$.*

The lemma relies on the following observation. Let $c_r(p) = B(r,k-r+1,p)$ and $d_r(p) = b(r,k-r+1,p)$ for convenience. If $p = 0.5$ then $c_r(p) = c_{k-r+1}(p)$, $d_r(p) = d_{k-r+1}(p)$ for all r.

Thus for any allocation vector $\boldsymbol{q}$ there is an allocation vector given by $\boldsymbol{q}^* = \frac{1}{2}(\boldsymbol{q}+\tilde{\boldsymbol{q}})$ such that $V(\boldsymbol{q},p) = V(\boldsymbol{q}^*,p)$. Hence the optimal allocation vector $\boldsymbol{q}^*$ can be made symmetric, i.e, the elements of $\boldsymbol{q}^*$ satisfy $q_r^* = q_{k-r+1}^*$.

Lemma 4.5. *If $p = 0.5$ and $\boldsymbol{q}$ is symmetric then*

$$s_{\boldsymbol{q}}(0.5) = \sum_{r=1}^k q_r F_{(r)}(x(0.5)) = 0.5.$$

To verify the lemma, first consider the case that F is symmetric. If F is symmetric about μ then, for any symmetric $\boldsymbol{q}$, $F_{\boldsymbol{q}}$ is also symmetric about μ. Note that, in this case, $\mu = x(0.5)$. Hence we have $s_{\boldsymbol{q}}(0.5) = 0.5$. But, since the quantities $F_{(r)}(x(0.5))$ do not depend on F, the lemma follows.

Table 4.8. Optimal unbalanced RSS designs for estimating a single quantile ξ_p, for selected p and set size k, based on minimizing asymptotic variance and assuming perfect ranking.

p \ k		3	4	5	6	7	8	9	10
0.05	$r(\boldsymbol{q}^*)$	1	1	1	1	1	1	1	1
	$s_{\boldsymbol{q}^*}(p)$	0.14	0.19	0.23	0.26	0.30	0.34	0.37	0.40
0.10	$r(\boldsymbol{q}^*)$	1	1	1	1	1	1	2	2
	$s_{\boldsymbol{q}^*}(p)$	0.27	0.34	0.41	0.47	0.52	0.57	0.23	0.26
0.15	$r(\boldsymbol{q}^*)$	1	1	1	2	2	2	2	2
	$s_{\boldsymbol{q}^*}(p)$	0.39	0.48	0.56	0.22	0.28	0.34	0.40	0.46
0.20	$r(\boldsymbol{q}^*)$	1	1	2	2	2	2	2	3
	$s_{\boldsymbol{q}^*}(p)$	0.49	0.59	0.26	0.34	0.42	0.50	0.56	0.32
0.25	$r(\boldsymbol{q}^*)$	1	2	2	2	2	3	3	3
	$s_{\boldsymbol{q}^*}(p)$	0.58	0.26	0.37	0.47	0.56	0.32	0.40	0.47
0.30	$r(\boldsymbol{q}^*)$	1	2	2	2	3	3	3	4
	$s_{\boldsymbol{q}^*}(p)$	0.66	0.35	0.47	0.58	0.35	0.45	0.54	0.35
0.35	$r(\boldsymbol{q}^*)$	2	2	2	3	3	3	4	4
	$s_{\boldsymbol{q}^*}(p)$	0.28	0.44	0.57	0.35	0.47	0.57	0.39	0.49
0.40	$r(\boldsymbol{q}^*)$	2	2	3	3	3	4	4	5
	$s_{\boldsymbol{q}^*}(p)$	0.35	0.52	0.32	0.46	0.58	0.41	0.52	0.37
0.45	$r(\boldsymbol{q}^*)$	2	2	3	3	4	4	5	5
	$s_{\boldsymbol{q}^*}(p)$	0.43	0.61	0.41	0.56	0.39	0.52	0.38	0.50

Lemma 4.6. *Let $\boldsymbol{q}$ be any allocation vector. Let $\tilde{\boldsymbol{q}}$ be the allocation vector obtained by reversing the components of $\boldsymbol{q}$. Then, for any $0 < p < 1$, $V(\boldsymbol{q}, p) = V(\tilde{\boldsymbol{q}}, 1-p)$, $s_{\boldsymbol{q}}(p) = 1 - s_{\tilde{\boldsymbol{q}}}(1-p)$.*

This lemma follows from the equalities: $c_r(p) = 1 - c_{k-r+1}(1-p)$,$d_r(p) = d_{k-r+1}(1-p)$. These can be easily verified by the definition of $c_r(p)$and $d_r(p)$.

Lemma 4.6 implies that if $\boldsymbol{q}^*$ is an optimal allocation vector and $s_{\boldsymbol{q}^*}(p)$ is the corresponding rank for the inference on the pth quantile, then $\tilde{\boldsymbol{q}}^*$ is an optimal allocation vector and $1 - s_{\boldsymbol{q}^*}(p)$ is the corresponding rank for the inference on the $(1-p)$th quantile. In other words, the optimal schemes have a symmetric structure. Hence, we only need to compute the optimal schemes for $0 < p \leq 0.5$.

The optimal schemes for $p = 0.05$ to 0.5 in steps of 0.05, $k = 3, \ldots, 10$, are computed. In all cases except for $p = 0.5$, optimal allocation vectors have only one non-zero component. In the case of $p = 0.5$, the allocations are equal on the medians of the sets of size k. That is, if k is odd then $q^*_{(k+1)/2} = 1$, and if k is even then $q^*_{k/2} = q^*_{k/2+1} = 0.5$. The index $r(\boldsymbol{q}^*)$ of the non-zero component of $\boldsymbol{q}^*$ and the corresponding $s_{\boldsymbol{q}^*}(p)$ of the optimal designs for $p = 0.05$ to 0.45 in steps of 0.05, $k = 3, \ldots, 10$, are given in Table 4.8.

4.5.2 Optimal RSS scheme for the simultaneous estimation of several quantiles

A more general result of asymptotic normality follows from the Bahadur representation of the unbalanced RSS sample quantiles.

Corollary 2 *Let* $0 < p_1 < \cdots < p_l < 1$ *be* l *probabilities. Let* $\hat{\boldsymbol{\xi}} = (\hat{z}_{\boldsymbol{q}_n}(p_1), \ldots, \hat{z}_{\boldsymbol{q}_n}(p_l))'$ *and* $\boldsymbol{\xi} = (z_{\boldsymbol{q}}(p_1), \ldots, z_{\boldsymbol{q}}(p_l))'$. *Then* $\sqrt{n}(\hat{\boldsymbol{\xi}} - \boldsymbol{\xi}) \to N(0, \Sigma)$, *where* Σ *is a positive definite matrix and, for* $i < j$, *the* (i, j)*th entry of* Σ *is*

$$\sigma_{ij} = \frac{\sum_{r=1}^{k} q_r F_{(r)}(z_{\boldsymbol{q}}(p_i))[1 - F_{(r)}(z_{\boldsymbol{q}}(p_j))]}{f_{\boldsymbol{q}}(z_{\boldsymbol{q}}(p_i)) f_{\boldsymbol{q}}(z_{\boldsymbol{q}}(p_j))}.$$

The Corollary above can be applied to determine optimal RSS schemes for a simultaneous estimation of several quantiles. Without loss of generality, the case of simultaneous estimation of two quantiles is considered in the following.

Let $s_1 = s_{\boldsymbol{q}}(p_1) = F_{\boldsymbol{q}}(x(p_1))$ and $s_2 = s_{\boldsymbol{q}}(p_2) = F_{\boldsymbol{q}}(x(p_2))$. Then the p_1th and p_2th quantiles, $x(p_1)$ and $x(p_2)$, of F are the s_1th and s_2th quantiles of $F_{\boldsymbol{q}}$ respectively. Note that $f_{\boldsymbol{q}}(x(p_1)) = \sum_{r=1}^{k} q_r d_r(p_1) f(x(p_1))$, $f_{\boldsymbol{q}}(x(p_2)) = \sum_{r=1}^{k} q_r d_r(p_2) f(x(p_2))$. Then it follows from the corollary that

$$\sqrt{n}\left[\begin{pmatrix} \hat{z}_{\boldsymbol{q}_n}(s_1) \\ \hat{z}_{\boldsymbol{q}_n}(s_2) \end{pmatrix} - \begin{pmatrix} x(p_1) \\ x(p_2) \end{pmatrix}\right] \to N\left(0, \Sigma(\boldsymbol{q})\right),$$

in distribution, where $\Sigma(\boldsymbol{q}) = C^{-1} B^{-1}(\boldsymbol{q}) A(\boldsymbol{q}) B^{-1}(\boldsymbol{q}) C^{-1}$, and

$$C = \begin{pmatrix} f(x(p_1)) & 0 \\ 0 & f(x(p_2)) \end{pmatrix},$$

$$B(\boldsymbol{q}) = \begin{pmatrix} \sum_{r=1}^{k} q_r d_r(p_1) & 0 \\ 0 & \sum_{r=1}^{k} q_r d_r(p_2) \end{pmatrix},$$

$$A(\boldsymbol{q}) = \begin{pmatrix} \sum_{r=1}^{k} q_r c_r(p_1)[1 - c_r(p_1)] & \sum_{r=1}^{k} q_r c_r(p_1)[1 - c_r(p_2)] \\ \sum_{r=1}^{k} q_r c_r(p_1)[1 - c_r(p_2)] & \sum_{r=1}^{k} q_r c_r(p_2)[1 - c_r(p_2)] \end{pmatrix}.$$

Let $V(\boldsymbol{q}) = B^{-1}(\boldsymbol{q}) A(\boldsymbol{q}) B^{-1}(\boldsymbol{q})$. Optimal RSS schemes can be determined based on $V(\boldsymbol{q})$. However, unlike the case of a single quantile, here we are faced with the choice of optimality criteria. Various criteria can be considered, such as D-optimality, A-optimality, etc. Suppose that the choice of criterion has been made and that the optimality criterion entails the minimization of a function of $V(\boldsymbol{q})$, say, $G(V(\boldsymbol{q}))$. Then the algorithm for determining optimal RSS schemes for the simultaneous estimation of several quantiles can be described as follows.

Step 1. Minimize $G(V(\boldsymbol{q}))$ with respect to $\boldsymbol{q}$ to determine the optimal allocation vector $\boldsymbol{q}^* = (q_1^*, \ldots, q_k^*)'$.

Step 2. Compute $s_j^* = \sum q_r^* B(r, k - r + 1, p_j)$ for $j = 1, \ldots, l$.

Table 4.9. Optimal unbalanced RSS designs for estimating a pair of quantiles (ξ_p, ξ_{1-p}), for selected p and set size k, based on minimizing asymptotic generalized variance and assuming perfect ranking.

p \ k		3	4	5
0.01	q^*	(0.5, 0, 0.5)	(0.5,0,0,0.5)	(0.5,0,0,0,0.5)
	$s_{q^*}(p)$	(0.015,0.985)	(0.020,0.980)	(0.025, 0.975)
0.05	q^*	(0.5, 0, 0.5)	(0.5,0,0,0.5)	(0.5,0,0,0,0.5)
	$s_{q^*}(p)$	(0.071,0.929)	(0.093,0.907)	(0.113, 0.887)
0.10	q^*	(0.5, 0, 0.5)	(0.5,0,0,0.5)	(0.5,0,0,0,0.5)
	$s_{q^*}(p)$	(0.136,0.864)	(0.172,0.828)	(0.205, 0.795)
0.15	q^*	(0.5, 0, 0.5)	(0.5,0,0,0.5)	(0.5,0,0,0,0.5)
	$s_{q^*}(p)$	(0.195,0.805)	(0.239,0.761)	(0.278, 0.722)
0.20	q^*	(0, 1, 0)	(0,0.5,0.5,0)	(0,0.5,0,0.5,0)
	$s_{q^*}(p)$	(0.104,0.896)	(0.104,0.896)	(0.135, 0.865)
0.25	q^*	(0, 1, 0)	(0,0.5,0.5,0)	(0, 0, 1, 0, 0)
	$s_{q^*}(p)$	(0.156,0.844)	(0.156,0.844)	(0.104, 0.896)

For illustration, consider the criterion of D-optimality in what follows. This entails the minimization of $|V(\boldsymbol{q})|$, the determinant of $V(\boldsymbol{q})$. As important examples, the D-optimal RSS schemes for the estimation of the pairs (ξ_p, ξ_{1-p}) for $p = 0.01, 0.05, 0.1, 0.15, 0.2$ and 0.25 are computed. The optimal allocation vectors and the corresponding vector $s_{\boldsymbol{q}^*}(\boldsymbol{p}) = (s_{\boldsymbol{q}^*}(p), s_{\boldsymbol{q}^*}(1-p))$ are given in Table 4.9.

4.6 The computation of optimal designs

In Section 4.4 and Section 4.5, we deliberately avoided the computational problem of the optimal designs. Now the problem is addressed.

The following lemma helps in developing algorithms for the computation.

Lemma 4.7. *Let $g(\mathbf{a})$ be a function on R^q whose gradient has no zero. Let $\mathcal{A} = \{\mathbf{a}_1, \cdots, \mathbf{a}_m\}$ be a set of non-zero q-vectors. Then the maximum (or minimum) of g on the convex hull spanned by the vectors in set $\mathcal{A}$ attains on a convex hull spanned by at most q vectors in set $\mathcal{A}$.*

The lemma follows from the Caratheodory Lemma and the fact that the gradient of $g(\mathbf{a})$ has no zero. The Caratheodory Lemma can be found in standard text books on nonlinear programming. It follows immediately from the lemma that

Corollary 3 *Let $g(A)$ be a function on $s \times s$ symmetric matrices whose gradient has no zeros. Let $\{A_1, \cdots, A_m\}$ be a set of non-zero $s \times s$ symmetric matrices. Set $h(\mathbf{q}) = g(A(\mathbf{q}))$ where $A(\mathbf{q}) = \sum_{i=1}^m q_i A_i$. Then, the maximum*

(or minimum) of h on the m-simplex $\{\boldsymbol{q} = (q_1, \cdots, q_m) : q_i \geq 0, \sum_{i=1}^{m} q_i = 1\}$ *attains at a point* $\boldsymbol{q}$ *that has at most* $\frac{1}{2}s(s+1)$ *non-zero components.*

The computation of all the optimal RSS schemes considered can be formulated in the framework of Corollary 3 and hence some simple algorithm can be developed for the computation. For example, according to Corollary 3, the minimum of the determinant or the trace of $[A(\boldsymbol{q})]^{-1}$ in Section 4.3 can be searched on three-dimensional simplexes which can, in turn, be reduced to two-dimensional triangles. The computation of the minimum can be done through a simple grid-point search algorithm.

4.7 Asymptotic relative efficiency of the optimal schemes

Let $\hat{\theta}_{n1}$ and $\hat{\theta}_{n2}$ be two estimators of θ. The asymptotic relative efficiency (ARE) of $\hat{\theta}_{n1}$ with respect to $\hat{\theta}_{n2}$ is defined as the ratio of the asymptotic generalized variance of $\hat{\theta}_{n2}$ and that of $\hat{\theta}_{n1}$. In this section, the ARE of the optimal RSS schemes of Section 4.4 and Section 4.5 are discussed.

4.7.1 Asymptotic relative efficiency of the optimal schemes for parametric location-scale families

The information matrix about $\boldsymbol{\theta}$ of a balanced RSS sample is, as derived in Section 3.1, given by

$$I(\boldsymbol{\theta}, \boldsymbol{q}_m) = mkI(\boldsymbol{\theta}) + mk(k-1)\Delta(\boldsymbol{\theta}),$$

where $\boldsymbol{q}_m = (m, \ldots, m)'$, and $I(\boldsymbol{\theta})$ is the information matrix of a single ordinary observation from distribution $F(x; \boldsymbol{\theta})$ and $\Delta(\boldsymbol{\theta})$ is a matrix with its (i,j)-th entry given by

$$\Delta_{ij}(\boldsymbol{\theta}) = E\left\{\frac{\frac{\partial F(X;\boldsymbol{\theta})}{\partial \theta_i}\frac{\partial F(X;\boldsymbol{\theta})}{\partial \theta_j}}{F(X;\boldsymbol{\theta})[1-F(X;\boldsymbol{\theta})]}\right\}.$$

As $n = km$ goes to infinity, the MLE of $\boldsymbol{\theta}$ converges in distribution to a normal distribution with mean $\boldsymbol{\theta}$ and variance-covariance matrix $[I(\boldsymbol{\theta}) + (k-1)\Delta(\boldsymbol{\theta})]^{-1}$. In the case of location-scale families, $I(\boldsymbol{\theta}) + (k-1)\Delta(\boldsymbol{\theta}) = [I_0 + (k-1)\Delta_0]/\sigma^2$, where I_0 and Δ_0 are the quantities corresponding to $\mu = 0, \sigma = 1$.

For a fixed set size k, let $\boldsymbol{q}_{ML}^{(k)}$ denote the optimal allocation proportions for the MLE. The asymptotic variance-covariance matrix of the optimal RSS MLE is $\sigma^2 A(\boldsymbol{q}_{ML}^{(k)})^{-1}$. Thus we have that the ARE of the optimal RSS MLE relative to the original balanced RSS MLE is given by

$$\text{ARE(OPT, ORI)} = \frac{|A(\boldsymbol{q}_{ML}^{(k)})|}{|I_0 + (k-1)\Delta_0|}.$$

Table 4.10. The relative efficiencies for the normal and extreme value distributions

set	Normal		Extreme value	
size	OPT/ORI	MLE/BLUE	OPT/ORI	MLE/BLUE
3	1.23	2.26	1.17	3.26
4	1.36	1.66	1.20	2.36
5	1.46	1.43	1.24	2.05
6	1.55	1.33	1.27	1.77
7	1.65	1.29	1.28	1.58
8	1.76	1.29	1.31	1.49
9	1.89	1.28	1.33	1.43
10	2.02	1.27	1.34	1.35

For $k = 3, \ldots, 10$, the ARE for the normal and extreme value families are presented in Table 4.10. Also given in the table is the ARE of the optimal RSS MLE relative to the optimal RSS BLUE.

The efficiency of the optimal RSS relative to the original RSS is quite substantial. The optimal RSS MLE is much more efficient than the optimal RSS BLUE for small set sizes. But the advantage of the MLE over the BLUE diminishes as the set size gets larger. Although the efficiency of the optimal RSS MLE relative to the optimal RSS BLUE is always bigger than one, the reader should not be misled against the BLUE by these numbers. The comparison is made only for the asymptotic variances. If sample sizes are small, the bias of the MLE will creep in. It might compromise the efficiency of the MLE over the BLUE. For small sample sizes, the optimal BLUE is a strong competitor of the optimal MLE, especially, when the set size is large.

4.7.2 Relative efficiency of the optimal schemes for the estimation of quantiles

For reasons that will become clear later, we consider the ARE of the optimal RSS schemes with respect to the SRS schemes. The SRS counterpart of the estimator of $x(p)$ is the pth sample quantile $\hat{\xi}_p$, asymptotically normal with mean $x(p)$ and variance $p(1-p)/[nf^2(x(p))]$. (See, e.g., Serfling [146], Chapter 2.) The ARE of the optimal RSS scheme with respect to the SRS scheme for estimating $x(p)$ is then given by

$$\text{ARE}(\hat{x}_{\boldsymbol{q}_n^*}(p), \hat{\xi}_p) = \frac{p(1-p)}{\sum_{r=1}^k q_r^* c_r(p)[1 - c_r(p)]/[\sum_{r=1}^k q_r^* d_r(p)]^2}.$$

The AREs of the optimal RSS scheme with respect to the SRS scheme, for $k = 2, \ldots, 9$ and $p = 0.05$ to 0.5 in steps of 0.05, are presented in Table 4.11. It can be seen that the gain in efficiency by using the optimal RSS schemes is large, the n quantified order statistics do almost as well as a simple random sample of size nk.

Table 4.11. The ARE of optimal unbalanced RSS schemes with respect to SRS schemes, for selected p and set size k, assuming perfect ranking.

p \ k	2	3	4	5	6	7	8	9
0.05	1.949	2.848	3.698	4.501	5.258	5.970	6.639	7.267
0.10	1.895	2.690	3.392	4.005	4.537	4.991	5.375	6.118
0.15	1.838	2.528	3.084	3.519	4.054	4.906	5.680	6.365
0.20	1.778	2.361	2.775	3.365	4.279	4.966	5.517	5.933
0.25	1.714	2.189	2.763	3.590	4.243	4.714	5.337	6.142
0.30	1.647	2.014	2.879	3.569	4.025	4.734	5.483	6.014
0.35	1.576	2.095	2.912	3.433	4.059	4.817	5.299	6.005
0.40	1.500	2.182	2.874	3.307	4.161	4.671	5.368	6.047
0.45	1.419	2.233	2.773	3.463	4.102	4.680	5.408	5.879
0.50	1.333	2.250	2.618	3.516	3.896	4.785	5.172	6.056

It is also interesting to compare the ARE of the optimal RSS schemes with the ARE of the balanced RSS schemes. As discussed in Section 2.6, the ARE of a balanced RSS scheme with respect to the corresponding SRS scheme is given by

$$\text{ARE}(\hat{x}_n(p), \hat{\xi}_p) = \frac{p(1-p)}{(1/k)\sum_{r=1}^{k} c_r(p)[1-c_r(p)]}.$$

As a function of p, the ARE of the balanced RSS schemes is a bow shaped curve with its maximum at $p = 0.5$ and reduces to 1 at both ends. Though the efficiency gain by using balanced RSS for the estimation of median is quite significant, the efficiency gain for the estimation of extreme quantiles is almost negligible. However, the optimal unbalanced RSS schemes achieve about the same efficiency gain for all quantiles. This is due to the effect that every quantile of the underlying distribution is made a central quantile of the $F_{\mathbf{q}}$ corresponding to the optimal unbalanced RSS scheme.

The optimal unbalanced RSS schemes for the estimation of quantiles bear a similarity to other statistical procedures such as importance sampling and saddlepoint approximation. They share the idea that if data values are sampled in a way which makes it more likely for a statistic to assume a value in the vicinity of a given point of interest, then that point may be estimated or approximated with greater accuracy. This idea can be explored further in the design of optimal RSS schemes for more complicated problems.

4.8 Other design methodologies

The optimal design for non-location-scale parametric families depends, in general, on unknown parameters. This is also the case for some other design problems. For example, it is well known that, for the nonparametric estimation of mean, the optimal design for unbiased estimators is given by the so-called Neyman allocation, that is, the proportion of allocation for the rth order statistic is proportional to its standard deviation. Specifically, $q_r = \sigma_{(r)}/(\sigma_{(1)} + \sigma_{(2)} + \cdots + \sigma_{(k)})$, where $\sigma_{(r)}$ is the standard deviation of the rth order statistic. When an optimal design depends on unknown parameters, the design can not be implemented without certain knowledge of the unknown parameters. The methodology of Bayes design or adaptive design is in order. In this section, the Bayes and adaptive design of RSS schemes are briefly discussed.

4.8.1 Bayesian design

Let $V(\boldsymbol{q}, \theta)$ denote a general optimality criterion where θ is an unknown parameter or vector of parameters and $\boldsymbol{q}$ is the variable allocation vector. Let $\pi(\theta)$ be a prior density function of θ that summarizes one's belief or certain previous experience. Let

$$B(\boldsymbol{q}) = \int V(\boldsymbol{q}, \theta)\pi(\theta)d\theta.$$

The Bayesian design amounts to the minimization of $B(\boldsymbol{q})$ with respect to $\boldsymbol{q}$. The challenge of Bayesian designs lies on the determination of an appropriate prior and the computation of $B(\boldsymbol{q})$.

4.8.2 Adaptive design

An adaptive design is a cyclical process as described below. At the initial step, a portion of the sample is taken according to any reasonable RSS scheme, say, the balanced RSS scheme. Then the process goes in cycles of an estimation stage and a design stage. In the estimation stage, the available data is used to estimate (or update the estimate of) θ. In the second stage, the updated estimate of θ is used to determine a tentative allocation vector $\boldsymbol{q}$ and more data are collected according to the tentative allocation vector. To assure the success of the adaptive design, a good initial estimate of θ is required. The following examples illustrate the method.

Example 1. The adaptive t-model. Kaur et al. [74] developed what they called the t-model for the optimal design of RSS for the estimation of mean of skew distributions. The model is motivated from the Neyman allocation and the fact that, in the case of skew distributions, the standard deviation of the largest order statistic is much larger than those of the other order statistics. The t-model specifies the following proportions of allocation:

$$q_1 = q_2 = \cdots = q_{k-1} = c, q_k = tc,$$

where $t > 1$ and $c = 1/(k - 1 + t)$. The optimal design amounts to finding the value of t such that the variance of the resultant unbiased estimator is the minimum. However, the optimal t value still depends on the unknown standard deviations of the order statistics as the Neyman allocation does, and the standard deviations of the order statistics can hardly be estimated. Instead of determining the optimal t value by a theoretical formula through the unknown standard deviations, Kaur et al. proposed a rule of thumb for the determination of the t value by an empirical relationship between the optimal t value and the coefficient of variation (CV) of the underlying distribution in the hope that, in field applications, one will somehow know the CV.

The adaptive method can be applied in conjunction with the t model when the CV has to be estimated from data. The procedure can start with a roughly estimated t value. Then at each step when data are obtained, the CV is estimated and the value of t updated. With the updated t value, the allocation scheme is modified and more data are taken. The procedure repeats until a predetermined sample size is reached.

Example 2. The adaptive estimation of mean through quantiles. The mean of a distribution can also be considered as a quantile, say, the pth quantile for some p. By the idea mentioned briefly at the end of Section 4.7, to estimate the mean more efficiently, the data should be drawn near this quantile. If p is known, the optimal design for the estimation of the quantile can be applied. Since p is unknown, an adaptive procedure needs to be carried out. Suppose an initial RSS sample (balanced or unbalanced) is already available. Based on the available sample, the method discussed in Section 4.2 is applied to estimate the mean and p. The estimated p is used to determine the RSS scheme for the sampling of the next step. Once new data is obtained, the estimate of p is updated. Then the procedure repeats itself.

4.9 Bibliographic notes

The analysis of general unbalanced RSS data was fully discussed in Chen [37]. The optimal RSS design for parametric families was developed in Chen and Bai [41]. The optimal RSS design for the estimation of quantiles was tackled in Chen [39]. The idea of nomination sampling was proposed by Willemain [170] independent of the idea of RSS. It was further considered by Tiwari [168] [169]. The lifetime data of r-out-of-k systems as unbalanced RSS data was treated by Kvam and Samaniego [85] and Chen [33]. Unbalanced RSS was considered by many authors in various contexts. Stokes [158] considered a special unbalanced RSS scheme for the estimation of the correlation coefficient in bivariate normal populations. Yu et al. [174] considered the estimation of variance by unbalanced RSS. Muttlak [103] [104] considered median RSS. Samawi et al. [141] considered the estimation of mean using extreme ranked set sampling. The first attempt at the design of optimal RSS was made by Stokes [161]. The optimal design for the estimation of mean was attempted by Kaur et al. [74] [70] and Ozturk and Wolfe [119]. Optimal RSS designs for distribution-free tests were considered by Ozturk [112] and Ozturk and

Wolfe [116]. Some details on optimal designs for distribution-free tests will be discussed in the next chapter.

5

Distribution-Free Tests with Ranked Set Sampling

In this chapter, the classical distribution-free tests: sign test, Mann-Whitney-Wilcoxon test and Wilconxon signed rank test, are studied in the context of RSS. For each of these tests, the RSS version of the test statistic is defined, the distributional properties of the test statistic are investigated, and the efficiency and power are compared with its counterpart in SRS. The related problems of parameter estimation including point estimates and confidence intervals are discussed. The effect of imperfect ranking in RSS on these tests is also addressed. The sign test is discussed in Section 5.1. The Mann-Whitney-Wilcoxon test is treated in Section 5.2. The Wilconxon signed rank test is dealt with in Section 5.3. In each of these three sections, the subsection dealing with the distributional properties of the test statistic is quite technical. Readers who are not interested in technicalities can skip these subsections. The last section of this chapter takes up the issues of optimal designs. Unbalanced RSS schemes which are optimal in terms of efficacy are considered for the sign test and Wilcoxon signed rank test.

5.1 Sign test with RSS

Let $H(x) = F(x - \theta)$ denote the distribution function of the underlying population, where F is a distribution function with a density function f and median 0. Consider the hypothesis:

$$H_0 : \theta = \theta_0 \quad \text{v.s.} \quad H_1 : \theta \neq \theta_0 \ \ (\text{or } > \theta_0, \ \text{ or } < \theta_0).$$

If we have a simple random sample $X_1, \cdots, X_N$, the classical sign test statistic on H_0 is given by

$$S^+_{\text{SRS}} = \sum_{j=1}^{N} I\{X_j - \theta_0 > 0\}.$$

In this chapter, sample sizes are denoted by capital letters. The test statistic S^+_{SRS} has a binomial distribution with parameters N and $p = P(X - \theta_0 > 0)$. The observed value of S^+_{SRS} should be stochastically close to, greater than or smaller than $N/2$ according as $\theta = \theta_0, \theta > \theta_0$ or $\theta < \theta_0$. This property indicates that H_0 should be rejected if the observed value of S^+_{SRS} is "far away" from $N/2$.

If, instead, we have a ranked-set sample $\{X_{[r]i}, r = 1, \cdots, k; i = 1, \cdots, m\}$, we define the RSS version of the sign test statistic as follows

$$S^+_{\text{RSS}} = \sum_{r=1}^{k}\sum_{i=1}^{m} I\{X_{[r]i} - \theta_0 > 0\}.$$

Throughout this chapter, we assume that the ranking mechanism in the RSS is consistent. In this section, we investigate the properties of S^+_{RSS} and its application to the hypothesis testing and estimation problems.

5.1.1 Distributional properties of S^+_{RSS} *

Let $N = mk$ and $h(x) = I\{x - \theta_0 > 0\}$. We can express S^+_{SRS} and S^+_{RSS}, respectively, by

$$S^+_{\text{SRS}} = N\bar{h}_{\text{SRS}} \quad \text{and} \quad S^+_{\text{RSS}} = N\bar{h}_{\text{RSS}},$$

where

$$\bar{h}_{\text{SRS}} = \frac{1}{N}\sum_{i=1}^{N} h(X_i) \;\text{ and }\; \bar{h}_{\text{RSS}} = \frac{1}{N}\sum_{r=1}^{k}\sum_{i=1}^{m} h(X_{[r]i}).$$

According to the results in Section 2.2 we have

$$E\bar{h}_{\text{SRS}} = E\bar{h}_{\text{RSS}} = Eh(X) \text{ and } \text{Var}(\bar{h}_{\text{RSS}}) \le \text{Var}(\bar{h}_{\text{SRS}}), \tag{5.1}$$

where the inequality is strict if there are at least two ranks, say r and s, such that $F_{[r]}(\theta_0 - \theta) \neq F_{[s]}(\theta_0 - \theta)$.

Since $Eh(X) = P(X > \theta_0) = \overline{H}(\theta_0) = \overline{F}(\theta_0 - \theta)$ where $\overline{H} = 1 - H$ and $\overline{F} = 1 - F$, we have

$$ES^+_{\text{RSS}} = N\overline{F}(\theta_0 - \theta). \tag{5.2}$$

Obviously, under the null hypothesis, $E(S^+_{\text{RSS}}) = \frac{1}{2}N$. When θ is close to θ_0, we have

$$E(S^+_{\text{RSS}}) \approx \frac{1}{2}N + Nf(0)(\theta - \theta_0). \tag{5.3}$$

The Variance of S^+_{RSS} can be derived as follows.

$$\begin{aligned}\text{Var}(S^+_{\text{RSS}}) &= m\sum_{r=1}^{k} Var(I(X_{[r]1} > \theta_0)) = m\sum_{r=1}^{k} \text{Var}(I(X_{[r]1} < \theta_0)) \\ &= m\sum_{r=1}^{k}[P(X_{[r]1} < \theta_0)) - P^2(X_{[r]1} < \theta_0))]\end{aligned}$$

$$= NF(\theta_0-\theta)\overline{F}(\theta_0-\theta) + m\sum_{r=1}^{k}[F^2(\theta_0-\theta) - F_{[r]}^2(\theta_0-\theta)]$$

$$= NF(\theta_0-\theta)\overline{F}(\theta_0-\theta) - m\sum_{r=1}^{k}[F_{[r]}(\theta_0-\theta) - F(\theta_0-\theta)]^2$$

$$= NF(\theta_0-\theta)\overline{F}(\theta_0-\theta)\delta^2,$$

where

$$\delta^2 = 1 - \frac{\sum_{r=1}^{k}[F_{[r]}(\theta_0-\theta) - F(\theta_0-\theta)]^2}{kF(\theta_0-\theta)\overline{F}(\theta_0-\theta)}, \tag{5.4}$$

and the fourth step follows from the fundamental equality

$$\sum_{r=1}^{k} F_{[r]} = kF.$$

Since $\mathrm{Var}(S_{\mathrm{SRS}}^+) = NF(\theta_0-\theta)\overline{F}(\theta_0-\theta)$, it follows from (5.1) that

$$0 < \delta^2 \leq 1, \tag{5.5}$$

where the second inequality is strict if there are at least two ranks, say r and s, such that $F_{[r]} \neq F_{[s]}$. The δ^2 is a measure of the reduction in variance or the gain in efficiency of RSS over SRS. It is in fact the inverse of what we have defined as the relative efficiency in the previous chapters. For this reason, we call δ^2 the variance reduction factor.

Under the null hypothesis, $\mathrm{Var}(S_{\mathrm{RSS}}^+) = \frac{1}{4}N\delta_0^2$, where δ_0 is the δ value under H_0, i.e., $\delta_0^2 = 1 - \frac{4}{k}\sum_{r=1}^{k}(F_{[r]}(0) - \frac{1}{2})^2$. When ranking is perfect in the RSS, $F_{[r]} = F_{(r)}$, and hence

$$F_{[r]}(0) = F_{(r)}(0) = B(r, k-r+1, F(0)) = B(r, k-r+1, 1/2),$$

where $B(r, s, x)$ is the distribution function of the beta distribution with parameters r and s. The values of $B(r, k-r+1, 1/2)$ can be obtained by using any standard statistical package. For example, it can be obtained in Splus by using the function *pbeta*. Note that, in the case of perfect ranking, δ_0^2 is free of F and completely known. We will see later that, in general, δ_0^2 is also free of F but it depends on the ranking mechanism and is not completely known.

The values of δ_0^2 with perfect ranking and $k = 2, \cdots, 9$, are given in Table 5.1. It can be seen that the reduction in variance by using RSS is substantial even for a small set size. With $k = 5$, the variance has been reduced to less than a half of the variance in SRS.

We now turn to the distribution of S_{RSS}^+. Write

$$S_{\mathrm{RSS}}^+ = \sum_{r=1}^{k} S_{m[r]} \tag{5.6}$$

Table 5.1. Values of $F_{(r)}(0)$, $r = 1, \cdots, k$ and δ_0^2

k:	2	3	4	5	6	7	8	9
1	0.750	0.875	0.938	0.969	0.984	0.992	0.996	0.998
2	0.250	0.500	0.688	0.813	0.891	0.938	0.965	0.981
3		0.125	0.313	0.500	0.656	0.773	0.856	0.910
4			0.063	0.188	0.344	0.500	0.637	0.746
5				0.031	0.109	0.227	0.363	0.500
6					0.016	0.063	0.145	0.254
7						0.008	0.035	0.090
8							0.004	0.020
9								0.002
δ_0^2	0.750	0.625	0.547	0.490	0.451	0.416	0.393	0.371

where

$$S_{m[r]} = \sum_{i=1}^{m} I\{X_{[r]i} - \theta_0 > 0\}.$$

We see that $S_{m[r]}$ follows a binomial distribution $B(m, p_r)$ with $p_r = \overline{F}_{[r]}(\theta_0 - \theta)$, and hence that S^+_{RSS} is the sum of k independent binomial variables. Under the null hypothesis, $p_r = p_r(0) = \overline{F}_{[r]}(0)$. If ranking in the RSS is perfect,

$$p_r(0) = \overline{F}_{(r)}(0) = 1 - B(r, k - r + 1, 1/2). \tag{5.7}$$

The remark on δ_0^2 applies to $p_r(0)$ as well. The $p_r(0)$'s are free of F, completely known when ranking is perfect and dependent of the ranking mechanism in general.

As a special case of the results in Section 2.2, we have the following asymptotic distribution of the S^+_{RSS}:

Theorem 5.1. *As $N \to \infty$,*

$$N^{-1/2}(S^+_{RSS} - N\overline{F}(\theta_0 - \theta)) \to N(0, F(\theta_0 - \theta)\overline{F}(\theta_0 - \theta)\delta^2).$$

Under the null hypothesis,

$$N^{-1/2}(S^+_{RSS} - \frac{1}{2}N) \to N(0, \frac{1}{4}\delta_0^2). \tag{5.8}$$

When N is large the asymptotic distribution of S^+_{RSS} in Theorem 5.1 can be used to determine the critical value for testing H_0. However, in the case of small samples, the exact distribution of S^+_{RSS} is preferred for more accuracy. We derive the exact null distribution of S^+_{RSS} in this sub-section. Recalling the decomposition (5.6), we have

$$P(S^+_{\text{RSS}} = y) = \sum_{i_1 + \cdots + i_k = y} \prod_{r=1}^{k} P(S_{m[r]} = i_r),$$

Table 5.2. Null Distribution of S^+_{RSS}: $P(S^+_{\text{RSS}} \geq y)$, $k = 2$

$y \backslash m$	5	6	7	8	9	10
0	1.0000	1.0000	1.0000			
1	.99977	.99996	.99999	1.0000	1.0000	
2	.99591	.99909	.99980	.99996	.99999	1.0000
3	.96900	.99158	.99784	.99947	.99987	.99997
4	.86772	.95505	.98614	.99602	.99891	.99972
5	.64510	.84497	.94126	.97992	.99365	.99811
6	.35490	.63243	.82576	.92802	.97321	.99080
7	.13228	.36757	.62262	.80933	.91550	.96622
8	.03100	.15503	.37738	.61473	.79509	.90374
9	.00409	.04495	.17424	.38527	.60819	.78262
10	.00023	.00842	.05874	.19067	.39181	.60266
11		.00091	.01386	.07198	.20491	.39734
12		.00004	00216	.02008	.08450	.21738
13			.00020	.00398	.02680	.09626
14			.00001	.00053	.00635	.03378
15				.00004	.00109	.00920
16					.00013	.00189
17					.00001	.00028
18						.00003

where

$$P(S_{m[r]}) = i) = \binom{m}{i} p_r^i (1 - p_r)^{m-i}$$

Thus, we establish the following theorem.

Theorem 5.2. *Under* H_0,

$$P(S^+_{RSS} = y) = \sum_{i_1 + \cdots + i_k = y} \prod_{r=1}^{k} \binom{m}{i_r} p_r(0)^{i_r} (1 - p_r(0))^{m - i_r}, \qquad (5.9)$$

where

$$p_r(0) = \overline{F}_{[r]}(0).$$

In the case of perfect ranking,

$$p_r(0) = 1 - B(r, k - r + 1, 1/2).$$

The exact distribution of S^+_{RSS} when ranking is perfect for $k \leq 4$ and $m \leq 10$ are given in Tables 5.2—5.4.

In the case of perfect ranking, the probability given in (5.9) can be rewritten as follows.

Table 5.3. Null Distribution of S^+_{RSS}: $P(S^+_{\text{RSS}} \geq y)$, $k = 3$

$y \setminus m$	3	4	5	6	10
0	1.00000	1.00000	.	.	.
1	.999836	.999991	1.00000	1.00000	.
2	.995841	.999700	.999980	.999999	.
3	.959311	.995850	.999635	.999971	.
4	.805446	.969380	.996343	.999629	.
5	.500000	.867672	.977264	.996965	1.00000
6	.194554	.644433	.907766	.983183	.999997
7	.040689	.355567	.745818	.934701	.999970
8	.004159	.132328	.500000	.817149	.999790
9	.000164	.030620	.254183	.618322	.998846
10		.004150	.092235	.381678	.994959
11		.000300	.022737	.182851	.982145
12		.000009	.003657	.065299	.948522
13			.000365	.016817	.877632
14			.000020	.003035	.757194
15				.000371	.591858
16				.000029	.408142
17				.000001	.242806
18					.122368
19					.051478
20					.017855
21					.005051
22					.001154
23					.000210
24					.000030
25					.000003

$$P(S^+_{\text{RSS}} = y) = \left[\binom{mk}{m, \cdots, m}(1/k)^{mk}\right]^{-1} \times$$

$$\sum_{i_1+\cdots+i_k=y} \binom{mk}{i_1, m-i_1, \cdots, i_k, m-i_k} \prod_{r=1}^{k} \left(\frac{p_r(0)}{k}\right)^{i_r} \left(\frac{1-p_r(0)}{k}\right)^{m-i_r}.$$

Here and in the sequel, for $a = a_1 + \cdots + a_k$, we use the notation

$$\binom{a}{a_1, \cdots, a_k} = \frac{a!}{a_1! \cdots a_k!}.$$

This expression can be interpreted by an urn model. Suppose that there are k urns each of which contains a certain number of white and red balls so that the probability with which a white ball is drawn from the r-th urn is $p_r(0)$. Let an urn be selected with equal probability, and then a ball be drawn from the selected urn and returned to the urn after recording its color. Repeat

Table 5.4. Null Distribution of S^+_{RSS}: $P(S^+_{\text{RSS}} \geq y)$, $k = 4$

$y \backslash m$	4	5	6	7	8	9	10
3	.999947	.999999	1.00000	.	.	.	.
4	.999162	.999977	.999999	.	.	.	.
5	.992350	.999688	.999990	1.00000	.	.	.
6	.957549	.997304	.999884	.999996	1.00000	.	.
7	.848648	.984393	.999032	.999957	.999998	.	.
8	.634270	.937472	.994236	.999648	.999984	1.00000	.
9	.365730	.820847	.974940	.997861	.999871	.999994	1.00000
10	.151352	.620178	.918681	.990121	.999203	.999952	.999998
11	.042451	.379822	.798605	.964777	.996143	.999702	.999982
12	.007650	.179153	.609767	.901507	.985198	.998503	.999888
13	.000838	.062528	.390233	.780344	.954411	.993905	.999421
14	.000053	.015607	.201395	.601667	.885933	.979699	.997526
15	.000002	.002696	.081319	.398333	.765027	.944144	.991208
16	.	.000312	.025060	.219656	.595132	.871826	.973815
17	.	.000023	.005764	.098493	.404878	.751948	.934150
18	.	.000001	.000968	.035223	.234973	.589715	.859023
19	.	.	.000116	.009879	.114067	.410285	.740618
20	.	.	.000010	.002139	.045589	.248052	.585129
21	.	.	.000001	.000352	.014802	.128174	.414871
22	.	.	.	.000043	.003857	.055856	.259382
23	.	.	.	.000004	.000797	.020301	.140977
24	.	.	.	.	.000129	.006095	.065850
25	.	.	.	.	.000016	.001497	.026185
26	.	.	.	.	.000002	.000298	.008792
27	.	.	.	.	.	.000048	.002474
28	.	.	.	.	.	.000006	.000579
29	.	.	.	.	.	.	.000112

this procedure mk times and denote by S the total number of white balls and by T_r the number of times that the urn r is selected. Then,

$$P(T_1 = \cdots = T_k = m) = \binom{mk}{m, \cdots, m} (1/k)^{mk}$$

and

$$P(S = y;\ T_1 = \cdots = T_k = m)$$
$$= \sum_{i_1 + \cdots + i_k = y} \binom{mk}{i_1, m - i_1, \cdots, i_k, m - i_k} \prod_{r=1}^{k} \left(\frac{p_r(0)}{k}\right)^{i_r} \left(\frac{1 - p_r(0)}{k}\right)^{m - i_r}.$$

Therefore,

$$P(S^+_{\text{RSS}} = y) = P(S = y | T_1 = \cdots = T_k = m).$$

5.1.2 Decision ruls of the sign test and confidence intervals for median

(i) One-sided test. To test $H_0 : \theta = \theta_0$ against the one-sided alternative $H_1 : \theta < \theta_0$, the null hypothesis is rejected if $S^+_{\text{RSS}} \leq \underline{m}_\alpha$, where $\underline{m}_\alpha$ is the largest integer such that $P(S^+_{\text{RSS}} \leq m) \leq \alpha$ and α is the significance level. To test $H_0 : \theta = \theta_0$ against the one-sided alternative $H_1 : \theta > \theta_0$, the null hypothesis is rejected if $S^+_{\text{RSS}} \geq \bar{m}_\alpha$, where $\bar{m}_\alpha$ is the smallest integer such that $P(S^+_{\text{RSS}} \geq m) \leq \alpha$.

(ii) Two-sided test. To test $H_0 : \theta = \theta_0$ against the two-sided alternative $H_1 : \theta \neq \theta_0$, the null hypothesis is rejected if $S^+_{\text{RSS}} < \underline{m}_{\alpha/2}$ or $S^+_{\text{RSS}} > \bar{m}_{\alpha/2}$, where $\underline{m}_{\alpha/2}$ and $\bar{m}_{\alpha/2}$ are determined in the same way as in the one-sided tests.

If N is small, the critical values $\underline{m}_{\alpha/2}$ and $\bar{m}_{\alpha/2}$ can be determined by using the exact distribution of S^+_{RSS}. They can be either read out from the Tables 5.2—5.4 or computed from formula (5.9). Otherwise, the critical values can be determined by the normal approximation. For example, $\underline{m}_\alpha \approx \frac{N}{2} - z_\alpha \sqrt{N\delta_0}/2 - 0.5$, where z_α is the $(1-\alpha)$th quantile of the standard normal distribution.

(iii) Confidence interval. To construct the confidence interval of θ, sort the ranked-set sample $\{X_{(r)i}, r = 1, \cdots, k; i = 1, \cdots, m\}$ as $X^*_{(1)} \leq \cdots \leq X^*_{(N)}$. Then, by using the normal approximation, an approximate confidence interval for θ of confidence level $1-\alpha$ is given by

$$(X^*_{([\frac{1}{2}N] - z_{\alpha/2}\sqrt{N\delta_0/2})}, X^*_{([\frac{1}{2}N]+1+z_{\alpha/2}\sqrt{N\delta_0/2})})$$

if N is even, and

$$(X^*_{([\frac{1}{2}N]+1-z_{\alpha/2}\sqrt{N\delta_0/2})}, X^*_{([\frac{1}{2}N]+1+z_{\alpha/2}\sqrt{N\delta_0/2})})$$

if N is odd. Note that subscripts appearing in the intervals should be round up to integers.

5.1.3 Effect of imperfect ranking on RSS sign test

We investigate the effect of imperfect ranking on the efficiency of RSS sign test in this sub-section. We consider any imperfect but consistent ranking mechanisms, e.g., judgment ranking with error or ranking with concomitant variables. We have derived in Section 3.1 that, if the ranking mechanism in RSS is consistent, the distribution function of the rth ranked order statistic can be expressed as

$$F_{[r]} = \sum_{s=1}^{k} p_{rs} F_{(s)},$$

where $p_{rs} \geq 0$, $\sum_{r=1}^{k} p_{rs} = 1 = \sum_{s=1}^{k} p_{rs}$ and $F_{(s)}$ is the distribution function of the actual sth order statistic. In fact, $p_{rs} = P(X_{[r]} = X_{(s)})$, the probability with which the rth ranked order statistic is the actual sth order statistic. We are going to consider the variance reduction factor δ^2. To distinguish, we denote the variance reduction factor by δ^2_{PFT} when ranking is perfect and by δ^2_{IPFT} when ranking is imperfect. Recall that the smaller the variance reduction factor, the more efficient the RSS.

We have already seen in (5.5) that

$$0 < \delta^2_{\text{IPFT}} < 1,$$

if not all the $F_{[r]}$'s are the same, which implies that an RSS, even if the ranking is imperfect, will always be more efficient than an SRS unless the ranking is a pure random permutation. However, an RSS with imperfect ranking will always be less efficient than the RSS with perfect ranking. We have

Theorem 5.3. *If $\theta_0 - \theta$ is inside the support of F, then*

$$\delta^2_{IPFT} \geq \delta^2_{PFT},$$

where the equality holds if and only if the consistent ranking is also perfect.

Proof. By Cauchy-Schwarz inequality

$$\begin{aligned} &\Big[\sum_{s=1}^{k} p_{rs}\{F_{(s)}(\theta_0-\theta) - F(\theta_0-\theta)\}\Big]^2 \\ &\leq \sum_{s=1}^{k} p_{rs}\Big[F_{(s)}(\theta_0-\theta) - F(\theta_0-\theta)\Big]^2 \end{aligned} \tag{5.10}$$

which implies that

$$\begin{aligned} &\sum_{r=1}^{k}\Big[F_{[r]}(\theta_0-\theta) - F(\theta_0-\theta)\Big]^2 \\ &= \sum_{r=1}^{k}\Big[\sum_{s=1}^{k} p_{rs}F_{(s)}(\theta_0-\theta) - F(\theta_0-\theta)\Big]^2 \\ &\leq \sum_{r=1}^{k}\sum_{s=1}^{k} p_{rs}\Big[F_{(s)}(\theta_0-\theta) - F(\theta_0-\theta)\Big]^2 \\ &= \sum_{s=1}^{k}\Big[F_{(s)}(\theta_0-\theta) - F(\theta_0-\theta)\Big]^2. \end{aligned} \tag{5.11}$$

Then it follows from (5.11) and the definition of the variance reduction factor that

$$\delta^2_{\text{IPFT}} \geq \delta^2_{\text{PFT}}.$$

Note that, the equality in (5.10) holds if and only if $p_{rs} = 1$ for one s, say $s = s(r)$, and $p_{rs} = 0$ for all other s. Then the equality in (5.11) holds if and only if

$$\sum_{r=1}^{k} F_{(r)} = \sum_{r=1}^{k} F_{(s(r))},$$

that is, $(s(1), \cdots, s(k))$ is a permutation of $(1, \cdots, k)$, which implies that, in every single cycle of the RSS, for each r, the actual rth order statistic is measured exactly once. Such an RSS is equivalent to the RSS with perfect ranking. The proof of the theorem is complete.

Let

$$\mathbf{1} = (1, \cdots, 1)'$$
$$\boldsymbol{F}^T(\theta) = (F_{(1)}(\theta_0 - \theta), \cdots, F_{(k)}(\theta_0 - \theta)),$$
$$\mathbf{P} = \begin{pmatrix} p_{11} \; p_{12} \cdots p_{1k} \\ p_{21} \; p_{22} \cdots p_{2k} \\ \cdots \cdots \cdots \cdots \\ p_{k1} \; p_{k2} \cdots p_{kk} \end{pmatrix}.$$

Then we can express δ^2_{IPFT} in a matrix form as follows.

$$\delta^2_{\text{IPFT}} = 1 - \frac{[\boldsymbol{F}(\theta) - F(\theta_0 - \theta)\mathbf{1}]^T \mathbf{P}\mathbf{P}^T [\boldsymbol{F}(\theta) - F(\theta_0 - \theta)\mathbf{1}]}{kF(\theta_0 - \theta)\bar{F}(\theta_0 - \theta)}.$$

Especially, under the null hypothesis,

$$\delta^2_{\text{IPFT}} = 1 - \frac{4}{k}[\boldsymbol{F}(\theta_0) - \mathbf{1}/2]^T \mathbf{P}\mathbf{P}^T [\boldsymbol{F}(\theta_0) - \mathbf{1}/2].$$

Note that

$$\boldsymbol{F}(\theta_0) = (F_{(1)}(0), \cdots, F_{(k)}(0))^T = (B(1, k, 1/2), \cdots, B(k, 1, 1/2))^T$$

which does not depend on F. Therefore, under the null hypothesis, δ^2_{IPFT} depends only on the ranking mechanism but not on the underlying distribution.

Now let us consider an example to see how the efficiency is affected by imperfect ranking. Let

$$\mathbf{P} = \begin{pmatrix} .75 & .25 & 0 & 0 & 0 \\ .25 & .50 & .25 & 0 & 0 \\ 0 & .25 & .50 & .25 & 0 \\ 0 & 0 & .25 & .50 & .25 \\ 0 & 0 & 0 & .25 & .75 \end{pmatrix}.$$

With the above $\mathbf{P}$, we have, under the null hypothesis, the variance reduction factor as 0.585. When ranking is perfect, the corresponding variance reduction factor is 0.490. This example shows that even if almost 50% of the rankings were wrong, the loss in variance reduction is only $[(1-0.490)-(1-0.585)]/[1-0.490] = 0.186$, that is, about 19%.

5.1.4 Comparison of efficiency and power with respect to S^+_{SRS}

We consider the Pitman's asymptotic relative efficiency of the RSS sign test with respect to the SRS sign test and make a comparison between the powers of the two tests by considering a number of underlying distribution families.

(i) Pitman's Asymptotic Relative Efficiency (ARE). We mentioned that the variance reduction factor δ^2 of S^+_{RSS} is the inverse of the relative efficiency of S^+_{RSS} with respect to S^+_{SRS} when these two statistics are considered as estimators of $EI\{X - \theta_0 > 0\}$. In fact, we can also show that the inverse of the efficiency coefficient δ_0^2 under the null hypothesis is the Pitman's ARE. That is,

$$\text{ARE}(S^+_{rss}, S^+_{srs}) = 1/\delta_0^2.$$

The ARE is always bigger than 1. By checking the values of δ_0^2 in Table 5.1, we can see that the RSS sign test is much more efficient than the SRS sign test.

(ii) Power Comparison. We first take up the small sample power comparison and consider the null hypothesis $H_0 : \theta = 0$ and the alternative hypothesis $H_1 : \theta = \Delta(> 0)$. We consider the RSS with $m = k = 5$. The significance level for the RSS sign test is taken as $\alpha_{\text{RSS}} = 0.0414$ and that for the SRS sign test is taken as $\alpha_{\text{SRS}} = 0.0539$. The corresponding decision rule for the RSS sign test is to reject H_0 if $S^+_{\text{RSS}} \geq 16$ and that for the SRS sign test is to reject H_0 if $S^+_{\text{SRS}} \geq 17$. The powers of the two tests are computed as

$$\beta_{\text{SRS}}(\Delta) = \sum_{y=17}^{25} \binom{25}{y} [1 - H(0)]^y [H(0)]^{25-y}$$

$$\beta_{\text{RSS}}(\Delta) = \sum_{y=16}^{25} \sum_{J_y} \prod_{r=1}^{5} \binom{5}{j_r} [p_r]^{j_r} [1 - p_r]^{5-j_r},$$

where $H(0) = F(-\Delta)$ is the distribution function under the alternative hypothesis evaluated at 0,

$$p_r = 1 - B(r, 6 - r, H(0)),$$

and the summation $\sum_{J_y}$ runs over all $(j_1, \cdots, j_5)$ such that $j_1 + \cdots + j_5 = y$. Since the power of the tests is not distribution free, we consider the following distribution families: (a) Double exponential distribution, (b) Cauchy distribution, and (c) Contaminated normal distribution.

Double Exponential Distribution. The density function of the double exponential distribution is

$$h(x) = \frac{1}{2} \exp(-|x - \theta|), \quad -\infty < x < \infty.$$

Under H_1, we have

Table 5.5. Power Comparison

(a) Double Exponential Distribution

Δ	0.0	0.1	0.25	0.5	0.75	1.0	1.25
β_{SRS}	.0539	.1288	.3108	.6638	.8865	.9713	.9941
β_{RSS}	.0414	.1475	.4469	.8767	.9876	.9993	1.000
β_{RSS}^P	.0563	.1691	.4515	.8545	.9792	.9980	.9998

(b) Cauchy Distribution

Δ	0.0	0.1	0.25	0.5	1.0	1.5	2.0
β_{SRS}	.0539	.0984	.2044	.4567	.8506	.9681	.9928
β_{RSS}	.0414	.1010	.2716	.6599	.9779	.9991	.9999
β_{RSS}^P	.0563	.1216	.2892	.6467	.9663	.9976	.9998

(c) Contaminated Normal Distribution

Δ	0.0	0.1	0.25	0.5	0.75	1.0	1.25
β_{SRS}	.0539	.1094	.2529	.6035	.8788	.9797	.9981
β_{RSS}	.0414	.1175	.3526	.8250	.9858	.9996	1.000
β_{RSS}^P	.0563	.1387	.3648	.8029	.9767	.9989	1.000

$$H(x) = \begin{cases} \frac{1}{2}\exp(x-\Delta) & \text{if } x \le \Delta, \\ 1-\frac{1}{2}\exp(\Delta - x) & \text{otherwise.} \end{cases}$$

Hence,

$$H(0) = \frac{1}{2}e^{-\Delta}.$$

Cauchy Distribution. The density function of the Cauchy distribution is

$$h(x) = \frac{1}{\pi}\frac{1}{1+(x-\theta)^2}, \quad -\infty < x < \infty.$$

Under H_1, we have

$$H(0) = \frac{1}{2} - \frac{1}{\pi}\arctan(\Delta).$$

Contaminated Normal Distribution. We consider the contaminated normal distribution with the following density function :

$$h(x) = 0.9\phi(x-\theta) + 0.1\phi((x-\theta)/2).$$

Under H_1, we have

$$H(0) = 0.9\Phi(-\Delta) + 0.1\Phi(-\Delta/2).$$

The powers for selected Δ values for the double exponential, Cauchy and contaminated normal distributions are presented, respectively, in Table 5.5(a)—(c) where β^P_{RSS} is the power for imperfect ranking with the matrix P given in section 5.1.3.

We now turn to the power comparison for large samples. Consider the root-N alternative $H_1 : \theta = \Delta/\sqrt{N}$, where $\Delta > 0$. When N is sufficiently large, we have

$$\begin{aligned} \beta_{\text{RSS}}(\Delta) &= P(S^+_{\text{RSS}} > \frac{1}{2}N + z_\alpha\sqrt{N}\delta_0/2 - 0.5) \\ &\approx P(Z > z_\alpha - 2\Delta f(0)/\delta_0) = \Phi(2\Delta f(0)/\delta_0 - z_\alpha), \end{aligned}$$

$Z \sim N(0,1)$ and Φ is its distribution function . As for the power of the RSS sign test with imperfect ranking, one only needs to replace the value of δ_0 by the corresponding value with imperfect ranking.

As is well known, the power function for the SRS sign test is

$$\beta_{\text{SRS}}(\Delta) \approx \Phi(2\Delta f(0) - z_\alpha).$$

Since $0 < \delta_0 < 1$, it is obvious that

$$\beta_{\text{RSS}}(\Delta) > \beta_{\text{SRS}}(\Delta).$$

5.2 Mann-Whitney-Wilcoxon test with RSS

In this section, we deal with the RSS version of the Mann-Whitney-Wilcoxon (MWW) test. Suppose X follows a distribution with distribution function F and density function f and Y follows a distribution with distribution function $G(\cdot) = F(\cdot - \Delta)$ and density function $g(\cdot) = f(\cdot - \Delta)$. The MWW test concerns the testing of the null hypothesis $H_0 : \Delta = 0$ against the alternative hypothesis $H_1 : \Delta \neq 0$ (or > 0 or < 0).

In the context of SRS, the MWW test is based on two independent simple random samples: $\{X_i, i = 1, \cdots, M\}$ from F and $\{Y_j, j = 1, \cdots, N\}$ from G. The MWW test statistic is defined by

$$W_{\text{SRS}} = \sum_{j=1}^{N} R_j,$$

where R_j is the rank of Y_j in the pooled sample $\{X_i, Y_j, i = 1, \cdots, M, j = 1, \cdots, N\}$. It can be verified that

$$W_{\text{SRS}} = \frac{1}{2}N(N+1) + U_{\text{SRS}},$$

where

$$U_{\mathrm{SRS}} = \sum_{i=1}^{M}\sum_{j=1}^{N} I\{X_i < Y_j\}.$$

The U_{SRS} above is in the form of a two-sample U-statistic. It is easy to deal with the properties of W_{SRS} through those of U_{SRS}. What follow are some properties of U_{SRS}. First we have

$$\begin{aligned} E(U_{\mathrm{SRS}}) &= MNP(X_1 < Y_1) \\ &= \frac{1}{2}MN + MN\int_0^{\Delta}\int_{-\infty}^{\infty} f(y)f(y-x)dydx. \end{aligned}$$

When Δ is small,

$$E(U_{\mathrm{SRS}}) \approx \frac{1}{2}MN + MN\Delta\int_{-\infty}^{\infty} f^2(y)dy. \tag{5.12}$$

Under the null hypothesis, we have

$$\begin{aligned} EU_{\mathrm{SRS}} &= \frac{1}{2}MN, \\ \mathrm{Var}(U_{\mathrm{SRS}}) &= \frac{1}{12}MN(M+N+1). \end{aligned}$$

Applying the central limit theorem for two-sample U-statistics, one can show that

$$\frac{U_{\mathrm{SRS}} - EU_{\mathrm{SRS}}}{\sqrt{\mathrm{Var}(U_{\mathrm{SRS}})}} \to N(0,1),$$

in distribution, under some minor conditions on the sample sizes M and N. Especially, under the null hypothesis,

$$\frac{U_{\mathrm{SRS}} - \frac{1}{2}MN}{\sqrt{\frac{1}{12}MN(M+N+1)}} \to N(0,1),$$

or, equivalently,

$$\sqrt{M+N}\left[\frac{U_{\mathrm{SRS}}}{MN} - \frac{1}{2}\right] \to N\left(0, \frac{1}{12\lambda(1-\lambda)}\right),$$

where $\lambda = \lim_{M+N\to\infty} M/(M+N)$. It follows from (5.12) that the efficacy[1] of the MWW test is

[1] If the statistic $T_n = T_n(X)$ satisfies the asymptotic normality that $\sqrt{n}(T_n(X) - \mu(\theta)) \Rightarrow N(0, \sigma^2(\theta))$ for each parameter value θ, then the Pitman efficacy of the statistic T_n at the parameter value θ is defined to be $Eff(T_n) = (\mu'(\theta))^2/\sigma^2(\theta)$, if μ is continuously differentiable and σ^2 is continuous.

$$\text{Eff}(U_{\text{SRS}}) = 12\lambda(1-\lambda)(\int f^2(y)dy)^2.$$

Now, let us consider the analogue of W_{SRS} in the context of RSS. Assume that $\{X_{(r)i}; r = 1, \cdots, k; i = 1, \cdots, m\}$ is an RSS sample from F and that $\{Y_{(s)j}; s = 1, \cdots, l; j = 1, \cdots, n\}$ is an RSS sample from G. We define the RSS version of the MWW test statistic as follows.

$$W_{\text{RSS}} = \sum_{s=1}^{l}\sum_{j=1}^{n} R_{sj},$$

where R_{sj} is the rank of $Y_{(s)j}$ in the pooled sample $\{X_{(r)i}, Y_{(s)j} : r = 1, \cdots, k; i = 1, \cdots, m; s = 1, \cdots, l; j = 1, \cdots, n\}$. As in the case of SRS, it can be shown that

$$W_{\text{RSS}} = \frac{1}{2}ln(ln+1) + U_{\text{RSS}},$$

where

$$U_{\text{RSS}} = \sum_{r=1}^{k}\sum_{i=1}^{m}\sum_{s=1}^{l}\sum_{j=1}^{n} I(X_{(r)i} < Y_{(s)j}).$$

For the comparison between W_{SRS} and W_{RSS}, let $M = km$ and $N = ln$ throughout this section. We discuss, in this section, the properties of W_{RSS}, its application to hypothesis testing and the construction of a confidence interval, and its relative efficiency with respect to W_{SRS}.

5.2.1 Distributional properties of U_{RSS}*

We first derive the mean and variance of U_{RSS}. For convenience of notation, we denote by the generic notation $X_{[r]}$ and $Y_{[s]}$ the ranked order statistics $X_{[r]i}$ and $Y_{[s]j}$ respectively. By the generic notation $\mathbf{X}$ and $\mathbf{Y}$, we denote a collection of X-related and Y-related random variables respectively. Their contents can be figured out from the context. By applying the fundamental equality, we can obtain

$$\begin{aligned}
E(U_{\text{RSS}}) &= mnE\sum_{r=1}^{k}\sum_{s=1}^{l} I\{X_{[r]} < Y_{[s]}\} \\
&= mnE[E(\sum_{r=1}^{k}\sum_{s=1}^{l} I\{X_{[r]} < Y_{[s]}\}|\mathbf{Y})] \\
&= mnE[E(k\sum_{s=1}^{l} I\{X < Y_{[s]}\}|\mathbf{Y})] \\
&= mnE[E(k\sum_{s=1}^{l} I\{X < Y_{[s]}\}|\mathbf{X})]
\end{aligned}$$

$$
\begin{aligned}
&= mnE[E(klI\{X < Y\}|\mathbf{X})] \\
&= mnklP(X < Y) \\
&= \frac{1}{2}MN + MN\int_0^{\Delta}\int_{-\infty}^{\infty} f(y)f(y-x)dydx.
\end{aligned}
$$

That is, $EU_{\text{RSS}} = EU_{\text{SRS}}$. Hence, (5.12) holds for U_{RSS} as well.

For the variance of U_{RSS}, we present the following result whose derivation is straightforward.

$$
\sigma^2_{U_{\text{RSS}}} = \text{Var}(U_{\text{RSS}}) = MN[N\zeta_{10} + M\zeta_{01} + \zeta_{11}],
$$

where

$$
\begin{aligned}
\zeta_{10} &= \frac{1}{kl^2}\sum_{r=1}^{k}\sum_{1\le s_1,s_2\le l} P(X_{[r]} < Y_{[s_1]1}, X_{[r]} < Y_{[s_2]2}) \\
&\quad - \frac{1}{kl^2}\sum_{r=1}^{k}\sum_{1\le s_1,s_2\le l} P(X_{[r]} < Y_{[s_1]})P(X_{[r]} < Y_{[s_2]}) \\
&= \frac{1}{k}\sum_{r=1}^{k}[P(X_{[r]} < Y_1, X_{(r)} < Y_2) - P(X_{[r]} < Y_1)P(X_{[r]} < Y_2)] \\
&= \frac{1}{k}\sum_{r=1}^{k}[EG^2(X_{[r]}) - (EG(X_{[r]}))^2] \\
&= \text{Var}(G(X)) - \frac{1}{k}\sum_{r=1}^{k}(EG(X_{(r)}) - EG(X))^2,
\end{aligned}
$$

$$
\begin{aligned}
\zeta_{01} &= \frac{1}{k^2 l}\sum_{1\le r_1,r_2\le k}\sum_{s=1}^{l} P(X_{[r_1]1} < Y_{[s]}, X_{[r_2]2} < Y_{[s]}) \\
&\quad - \frac{1}{k^2 l}\sum_{1\le r_1,r_2\le k}\sum_{s=1}^{l} P(X_{(r_1)} < Y_{(s)})P(X_{(r_2)} < Y_{(s)}) \\
&= \frac{1}{l}\sum_{s=1}^{l} P(X_1 < Y_{[s]}, X_2 < Y_{[s]}) - P(X < Y_{[s]})P(X < Y_{[s]}) \\
&= \frac{1}{l}\sum_{s=1}^{l}[EF^2(Y_{[s]}) - E^2F(Y_{[s]})] \\
&= \text{Var}(F(Y)) - \frac{1}{l}\sum_{s=1}^{l}[EF(Y_{[s]}) - EF(Y)]^2,
\end{aligned}
$$

$$\zeta_{11} = \frac{1}{kl}\left[\sum_{r=1}^{k}\sum_{s=1}^{l} P(X_{[r]} < Y_{[s]})P(X_{[r]} > Y_{[s]})\right.$$
$$- \sum_{r=1}^{k}\sum_{s=1}^{l} P(X_{[r]} < Y_{[s]1}, X_{[r]} < Y_{[s]2})$$
$$+ \sum_{r=1}^{k}\sum_{s=1}^{l} P(X_{[r]} < Y_{[s]})P(X_{[r]} < Y_{[s]})$$
$$- \sum_{r=1}^{k}\sum_{s=1}^{l} P(X_{[r]1} < Y_{[s]}, X_{[r]2} < Y_{[s]})$$
$$\left.+ \sum_{r=1}^{k}\sum_{s=1}^{l} P(X_{[r]} < Y_{[s]})P(X_{[r]} < Y_{[s]})]\right]$$
$$= \frac{1}{kl}\sum_{r=1}^{k}\sum_{s=1}^{l}\Big[\mathrm{Var}(I(X_{[r]} > Y_{[s]})) - \mathrm{Var}(G_{[s]}(X_{[r]}) - \mathrm{Var}(F_{[r]}(Y_{[s]})\Big].$$

Note that, in the derivation of the mean and variance above, only the consistency of the ranking mechanism is assumed. In the following, we further assume that the ranking is perfect. Then, under H_0, we have

$$\zeta_{10}^{0} = \frac{1}{k}\sum_{r=1}^{k}\frac{r(k-r+1)}{(k+1)^2(k+2)} = \frac{1}{6(k+1)},$$
$$\zeta_{01}^{0} = \frac{1}{l}\sum_{s=1}^{k}\frac{s(k-s+1)}{(k+1)^2(k+2)} = \frac{1}{6(l+1)},$$
$$\zeta_{11}^{0} = \frac{1}{2} - \frac{1}{kl}\sum_{r=1}^{k}\sum_{s=1}^{l}\eta_{r:s}^2 - \sum_{r=1}^{k}\sum_{s=1}^{l}[\eta_{r:ss} - \eta_{r:s}^2 + \eta_{rr:s} - \eta_{r:s}^2]$$
$$= \frac{1}{4} - \frac{1}{kl}\sum_{r=1}^{k}\sum_{s=1}^{l}\Big[(\frac{1}{2} - \eta_{r:s})^2 + [\eta_{r:ss} - \eta_{r:s}^2] + [\eta_{rr:s} - \eta_{r:s}^2]\Big],$$

where $\eta_{r:s}$, $\eta_{r:ss}$ and $\eta_{rr:s}$ are probabilities which can be computed as follows.

$$\eta_{r:s} = P(X_{(r)} < Y_{(s)})$$
$$= \frac{k!l!}{(r-1)!(k-r)!(s-1)!(l-s)!} \times$$
$$\iint_{0<u<v<1} u^{r-1}(1-u)^{k-r}v^{s-1}(1-v)^{l-s}dudv$$
$$= \sum_{t=0}^{k-r}\frac{k!l!(k+s-t-1)!(l-s+t)!}{(k+l)!(s-1)!(l-s)!t!(k-t)!}$$

$$= \sum_{t=0}^{k-r} \binom{k+s-t-1}{s-1} \binom{l-s+t}{t} \Big/ \binom{k+l}{k},$$

$$\eta_{r:ss} = P(X_{(r)} < Y_{(s)1}, X_{(r)} < Y_{(s)2})$$
$$= \frac{k!(l!)^2}{(r-1)!(k-r)![(s-1)!(l-s)!]^2} \times$$
$$\iint_{0<v,w<u<1} u^{k-r}(1-u)^{r-1}(vw)^{l-s}[(1-v)(1-w)]^{s-1} du dv dw$$
$$= \sum_{t_1,t_2=0}^{s-1} \binom{k+2l-r-t_1-t_2}{k-r, l-t_1, l-t_2} \binom{r+t_1+t_2-1}{r-1, t_1, t_2} \Big/ \binom{k+2l}{k, l, l},$$

$$\eta_{rr:s} = P(X_{(r)1} < Y_{(s)}, X_{(r)2} < Y_{(s)})$$
$$= \frac{(k!)^2 l!}{[(r-1)!(k-r)!]^2 (s-1)!(l-s)!} \times$$
$$\iint_{0<u,v<w<1} (uv)^{r-1}[(1-u)(1-v)]^{k-r} w^{s-1}(1-w)^{l-s} du dv dw$$
$$= \sum_{t_1,t_2=0}^{k-r} \binom{2k+s-t_1-t_2-1}{k-t_1, k-t_2, s-1} \binom{l-s+t_1+t_2}{l-s, t_1, t_2} \Big/ \binom{k+2l}{k, l, l}.$$

Note that for each pair (r, s),

$$\eta_{rr:s} - \eta_{rs}^2 = \mathrm{Var}(F_{(r)}(Y_{(s)})) \geq 0,$$
$$\eta_{r:ss} - \eta_{rs}^2 = \mathrm{Var}(G_{(s)}(X_{(r)})) \geq 0.$$

In computation, we can use the following facts:

(a) $\sum_{r=1}^{k} \eta_{r:ss}$

$$= \frac{k(l!)^2}{[(s-1)!(l-s)!]^2} \iint_{0<v,w<u<1} (vw)^{l-s}[(1-v)(1-w)]^{s-1} du dv dw$$
$$= \frac{k}{2l+1} \sum_{t_1,t_2=0}^{s-1} \binom{2l-t_1-t_2}{l-t_1} \binom{t_1+t_2}{t_1} \Big/ \binom{2l}{l},$$

(b) $\sum_{s=1}^{l} \eta_{rr:s}$

$$= \frac{(k!)^2 l}{[(r-1)!(k-r)!]^2} \iint_{0<u,v<w<1} (uv)^{r-1}[(1-u)(1-v)]^{k-r} du dv dw$$
$$= \frac{l}{2k+1} \sum_{t_1,t_2=0}^{k-r} \binom{2k-t_1-t_2}{k-t_1} \binom{t_1+t_2}{t_1} \Big/ \binom{2k}{k}.$$

The following result gives the asymptotic distribution of U_{RSS}.

Theorem 5.4. *Let* $\lambda = \lim_{M+N\to\infty} M/(M+N)$. *If the ranking mechanisms involved in the two-sample RSS are consistent and* $0 < \lambda < 1$ *then*

$$\sqrt{N+M}(U_{RSS} - EU_{RSS})/MN \to N(0, \sigma_U^2),$$

in distribution, where

$$\sigma_U^2 = \frac{\zeta_{10}}{\lambda} + \frac{\zeta_{01}}{1-\lambda},$$

provided that $\sigma_U^2 > 0$. *Further, if rankings are perfect then, under* H_0,

$$\sqrt{N+M}(U_{RSS}/MN - \frac{1}{2}) \to N(0, \sigma_{U0}^2),$$

in distribution, where

$$\sigma_{U0}^2 = \frac{1}{6\lambda(k+1)} + \frac{1}{6(1-\lambda)(l+1)}.$$

The theorem can be proved by the projection approach as employed in the central limit theorem for the two sample U-statistic. Note that when $k = l = 1$, Theorem 5.4 coincides with the result for the SRS version of the MWW statistic.

Under the null hypothesis and the assumption of perfect ranking, we show below that the distribution of U_{RSS} is independent of the underlying distribution and symmetric.

First consider the distribution-free property. Under H_0, $F = G$ and hence both $F(X_{(r)})$ and $F(Y_{(s)})$ can be considered as the order statistics of simple random samples from the uniform distribution $U(0,1)$. Let $\{U_{(r)i} : r = 1, \cdots, k; i = 1, \cdots, m\}$ and $\{V_{(s)j} : s = 1, \cdots, l; j = 1, \cdots, n\}$ denote two ranked set samples from $U(0,1)$. Then we have

$$\begin{aligned} U_{\text{RSS}} &= \sum_{r=1}^{k}\sum_{i=1}^{m}\sum_{s=1}^{l}\sum_{j=1}^{n} I\{F(X_{(r)i}) < F(Y_{(s)j})\} \\ &\stackrel{\mathcal{D}}{=} \sum_{r=1}^{k}\sum_{i=1}^{m}\sum_{s=1}^{l}\sum_{j=1}^{n} I\{U_{(r)i} < V_{(s)j}\}. \end{aligned}$$

Therefore, the distribution of U_{RSS} is independent of the underlying distribution.

For the symmetry property of the distribution of U_{RSS}, note that $U_{(r)i} \stackrel{\mathcal{D}}{=} 1 - U_{(k-r+1)i}$ for any r and i and that $V_{(s)j} \stackrel{\mathcal{D}}{=} 1 - V_{(l-s+1)j}$ for any s and j. Hence,

$$U_{\text{RSS}} \stackrel{\mathcal{D}}{=} \sum_{r=1}^{k}\sum_{i=1}^{m}\sum_{s=1}^{l}\sum_{j=1}^{n} I\{1 - U_{(k-r+1)i} < 1 - V_{(l-s+1)j}\}$$

$$= \sum_{r=1}^{k}\sum_{i=1}^{m}\sum_{s=1}^{l}\sum_{j=1}^{n} I\{U_{(r)i} > V_{(s)j}\}$$
$$= mnkl - U_{\text{RSS}},$$

which implies that the distribution of U_{RSS} under the null hypothesis is symmetric about $klmn/2$.

Finally, let us consider the efficacy of U_{RSS} and its ARE relative to U_{SRS}. By (5.12) and the fact that $EU_{\text{RSS}} = EU_{\text{SRS}}$, we derive from Theorem 5.4 that the efficacy of U_{RSS} is

$$\text{Eff}(U_{\text{RSS}}) = \frac{6\lambda(1-\lambda)(k+1)(l+1)(\int f^2(x)dx)^2}{\lambda l + (1-\lambda)k + 1}.$$

Hence, the ARE of U_{RSS} relative to U_{SRS} is given by

$$\text{ARE}(U_{\text{RSS}}, U_{\text{SRS}}) = \frac{(k+1)(l+1)}{2[\lambda l + (1-\lambda)k + 1]}.$$

Note that $\text{ARE}(U_{\text{RSS}}, U_{\text{SRS}}) > 1$.

5.2.2 Decision rules of the RSS Mann-Whitney-Wilcoxon test

(i) Large sample test. If $M + N$ is large, the null distribution of U_{RSS} can be well approximated by the normal distribution given in Theorem 5.4. The decision rules can be determined in the usual way. For example, to test H_0 against $H_1 : \Delta > 0$, the decision rule is to reject H_0 if

$$U_{\text{RSS}} > \frac{MN}{2} + z_\alpha \frac{MN}{\sqrt{M+N}} \sqrt{\frac{1}{6\lambda(k+1)} + \frac{1}{6(1-\lambda)(l+1)}} - \frac{1}{2}.$$

(ii) A modification when sample size is moderate. If sample size is moderate, it is better to use the variance of U_{RSS} rather than its asymptotic variance in the normal approximation. The decision rule for the test above is then to reject H_0 if

$$U_{\text{RSS}} > \frac{MN}{2} + z_\alpha \frac{MN}{\sqrt{M+N}} \times$$
$$\sqrt{\frac{1}{6\lambda(k+1)} + \frac{1}{6(1-\lambda)(l+1)} + \frac{\zeta_{11}^0}{\lambda(1-\lambda)(M+N)}} - \frac{1}{2}.$$

(iii) Small sample test. If sample size is small, the normal approximation is no longer accurate enough. The exact null distribution of U_{RSS} must be used to determine the decision rules. However, the computation of the null distribution is very complicated. For example, Bohn and Wolfe [27] considered the simplest case that $m = n = 1$ and $k = l = 2$. The observations $x_{(1)}, x_{(2)}, y_{(1)}, y_{(2)}$

Table 5.6. The Null Distribution of U_{RSS} with $m = n = 1$, $k = l = 2$

U_{RSS}	0	1	2	3	4
Prob.	1/10	17/90	19/45	17/90	1/10

have $4! = 24$ possible permutations. As an example, the probability for the permutation $(y_{(1)}, y_{(2)}, x_{(1)}, x_{(2)})$ can be computed as

$$16 \int\int_{0<x<y<z<w<1} (1-x)y(1-z)w dx dy dz dw$$
$$= 16[\frac{1}{72} - \frac{1}{280} - \frac{1}{105} + \frac{1}{384}] = \frac{137}{2520}.$$

This corresponds to one of 4 cases where $U_{\text{RSS}} = 0$. The null distribution of U_{RSS} is given in Table 5.6

5.2.3 The estimate and confidence interval of Δ

By assumption, $Y - \Delta$ has the same distribution as X. Therefore, the median of $Y - X - \Delta$ is zero or, equivalently, the median of $Y - X$ is Δ. Now define

$$\hat{F}_{Y-X}(t) = \frac{1}{klmn} \sum_{r=1}^{k} \sum_{s=1}^{s} \sum_{i=1}^{m} \sum_{j=1}^{n} I\{Y_{(s)j} - X_{(r)i} \le t\}.$$

Similar to the derivation of EU_{RSS}, we can obtain

$$E\hat{F}_{Y-X}(t) = P(Y - X \le t),$$

that is, $\hat{F}_{Y-X}(t)$ provides an unbiased estimate for $P(Y - X \le t)$, the distribution function of $Y - X$. Intuitively, a quantile of the empirical distribution $\hat{F}_{Y-X}(t)$ will provide a reasonable estimate for the corresponding quantile of the distribution of $Y - X$. In particular, the median of $\hat{F}_{Y-X}(t)$ will provide a reasonable estimate for Δ, the median of $Y - X$. In fact, by the technique used for dealing with two-sample U-statistics, we can derive the asymptotic properties of the quantiles of $\hat{F}_{Y-X}(t)$ such as strong consistency and asymptotic normality similar to those for the one-sample RSS quantiles given in Section 2.5. Therefore, Δ can be estimated by the sample median of the differences $\{Y_{(s)j} - X_{(r)i} : s = 1, \cdots, l; j = 1, \cdots, n; r = 1, \cdots, k; i = 1, \cdots, m\}$. Denote the order statistics of the differences by $\{D_{(\nu)} : \nu = 1, \cdots, mnkl\}$. Then the estimate of Δ is given by

$$\hat{\Delta} = \begin{cases} D_{(\frac{mnkl+1}{2})} & \text{if } mnkl \text{ is odd,} \\ \frac{1}{2}[D_{(\frac{mnkl}{2})} + D_{(\frac{mnkl}{2}+1)}], & \text{otherwise.} \end{cases}$$

The $100(1-\alpha)\%$ confidence interval of Δ is given by

$$[D_{(c)},\ \ D_{(mnkl-c+1)}],$$

where c is the maximum integer such that $P(U_{\text{RSS}} \leq c) \leq \alpha/2$.

5.2.4 Effect of imperfect ranking on RSS Mann-Whitney-Wilcoxon test

Comparing the expressions of ζ_{01}, ζ_{10} and ζ_{11} with their counterparts in the variance of U_{SRS}, we can conclude that, even if the rankings involved in the two-sample RSS are not perfect, the RSS version of the MWW test is always more efficient than the SRS version of the MWW test as long as the rankings are consistent. On the other hand, we can also show that imperfect consistent RSS is always less efficient than perfect RSS. The only problem with the imperfect RSS is that the null distribution cannot be completely determined without knowing the ranking error probabilities, even if the sample size is large. One remedy for this is to use the bootstrap method to obtain an approximation to the distribution of the estimator $\hat{\Delta}$ by resampling from each of the sub-samples $\{X_{[r]i}, i = 1, \cdots, m\}, r = 1, \cdots, k$, and $\{Y_{[s]j}, j = 1, \cdots, n\}, s = 1, \cdots, l$, and then construct a bootstrap confidence interval or confidence bound for Δ. The decision rules for the tests are then determined by the duality between the critical regions of the tests and the confidence intervals or bounds.

5.3 Wilcoxon signed rank test with RSS

Suppose that the underlying distribution is symmetric with distribution function $F(\cdot - \theta)$ where F is a continuous distribution function symmetric about 0. Let f denote the density function corresponding to F. The Wilcoxon signed rank test concerns the testing of the null hypotheses $H_0 : \theta = \theta_0$ against the alternative hypothesis $H_1 : \theta \neq \theta_0$ (or $> \theta_0$ or $< \theta_0$). In the context of SRS, the Wilcoxon signed rank test statistic is defined as

$$W^+_{\text{SRS}} = \sum_{i=1}^{N} R_i^+ I\{X_i - \theta_0 > 0\},$$

where $\{X_i : i = 1, \cdots, N\}$ is a simple random sample from the population, R_i^+ is the rank of $|X_i - \theta_0|$ among $\{|X_1 - \theta_0|, \cdots, |X_N - \theta_0|\}$. An alternative expression of W^+_{SRS} is given by

$$W^+_{\text{SRS}} = \sum_{i \leq j} I\left\{\frac{X_i + X_j}{2} > \theta_0\right\}.$$

The analogue of W^+_{SRS} in the context of RSS is as follows. Let $\{X_{(r)i} : r = 1, \cdots, k; i = 1, \cdots, n\}$ be a ranked set sample. The RSS version of the Wilcoxon signed rank test statistic is defined by

$$W^+_{\text{RSS}} = \sum_{\{(r_1,i_1),(r_2,i_2)\}} I\left\{\frac{X_{(r_1)i_1} + X_{(r_2),i_2}}{2} > \theta_0\right\},$$

where, in the summation $\sum_{\{(r_1,i_1),(r_2,i_2)\}}$, r_1 and r_2 run through $1,\cdots,k$, i_1 and i_2 run through $1,\cdots,n$ subject to the constraint that either $i_1 < i_2$ or $r_1 \le r_2$ while $i_1 = i_2$.

We deal with the properties of W^+_{RSS} and its applications in this section.

5.3.1 Distributional properties of W^+_{RSS} *

Denote by $\mu(\theta)$ and $V(\theta)$ the mean and variance of W^+_{RSS} respectively. Let $F_2 = F * F$ and $F_{(r)2} = F_{(r)} * F_{(r)}$ where $*$ denotes the convolution operation. We first derive the mean of W^+_{RSS}.

$$\begin{aligned}
\mu(\theta) &= \frac{n(n-1)}{2}\sum_{r_1=1}^{k}\sum_{r_2=1}^{k} P(X_{(r_1)1} + X_{(r_2)2} > 2\theta_0) \\
&\quad + n\sum_{r_1 \le r_2} P(X_{(r_1)} + X_{(r_2)} > 2\theta_0) \\
&= \frac{k^2n^2}{2}P(X_1 + X_2 > 2\theta_0) + nkP(X > \theta_0) \\
&\quad - \frac{n}{2}\sum_{r=1}^{k} P(X_{(r)1} + X_{(r)2} > 2\theta_0) \\
&= \frac{k^2n^2}{2}\bar{F}_2(2(\theta_0 - \theta)) + nk\bar{F}(\theta_0 - \theta) - \frac{n}{2}\sum_{r=1}^{k}\bar{F}_{(r)2}(2(\theta_0 - \theta)).
\end{aligned}$$

By using the fact that $F_{(r)2}(0) + F_{(k-r+1)2}(0) = 1$ which follows from

$$\begin{aligned}
F_{(r)2}(0) &= \int_{-\infty}^{0}\int_{-\infty}^{\infty} f_{(r)}(x)f_{(r)}(y-x)dxdy \\
&= \int_{0}^{\infty}\int_{-\infty}^{\infty} f_{(r)}(-u)f_{(r)}(u-v)dudv \\
&= \int_{0}^{\infty}\int_{-\infty}^{\infty} f_{(k-r+1)}(u)f_{(k-r+1)}(v-u)dudv \\
&= 1 - F_{(k-r+1)2}(0),
\end{aligned}$$

we obtain, under the null hypothesis, that

$$\mu(\theta_0) = \frac{nk(nk+1)}{4}.$$

When $\theta - \theta_0$ is small, we have

$$\mu(\theta) = \frac{1}{4}N(N+1) + N^2\Psi_{nk}(\theta - \theta_0) + o(\theta - \theta_0) \tag{5.13}$$

where

$$\begin{aligned}\Psi_{nk} &= f_2(0) + \frac{f(0)}{nk} - \frac{1}{nk^2}\sum_{r=1}^{k} f_{(r)2}(0) \\ &\to f_2(0) = \int f^2(y)dy,\end{aligned}$$

where f_2 and $f_{(r)2}$ are the density functions of F_2 and $F_{(r)2}$ respectively.

The variance of W^+_{RSS} is derived as follows.

$$V(\theta) = \sum_{\{(r_1,i_1),(r_2,i_2)\}} \sum_{\{(r_3,i_3),(r_4,i_4)\}} C([(r_1,i_1),(r_2,i_2)],[(r_3,i_3),(r_4,i_4)]), \tag{5.14}$$

where

$$\begin{aligned}&C([(r_1,i_1),(r_2,i_2)],[(r_3,i_3),(r_4,i_4)]) \\ &= \text{Cov}\left(I\left\{\frac{X_{(r_1)i_1}+X_{(r_2)i_2}}{2} > \theta_0\right\}, I\left\{\frac{X_{(r_3)i_3}+X_{(r_4)i_4}}{2} > \theta_0\right\}\right).\end{aligned}$$

If neither of (r_1,i_1) and (r_2,i_2) is the same as either (r_3,i_3) or (r_4,i_4) then $C([(r_1,i_1),(r_2,i_2)],[(r_3,i_3),(r_4,i_4)]) = 0$ since the two indicators are independent. The non-zero terms can be distinguished as follows:

1. $\{(r_1,i_1),(r_2,i_2)\} = \{(r_3,i_3),(r_4,i_4)\}$.
2. Exactly one of (r_1,i_1) and (r_2,i_2) is the same as either (r_3,i_3) or (r_4,i_4).

Case 2 can be further distinguished as

2a. There are three distinct i-indices.
2b. There are two distinct i-indices.
2c. All the i-indices are the same.

Let $V_1(\theta), V_{2a}(\theta), V_{2b}(\theta)$ and $V_{2c}(\theta)$ denote, respectively, the sum of the terms in Case 1, 2a, 2b and 2c. Hence we can write

$$V(\theta) = V_1(\theta) + V_{2a}(\theta) + V_{2b}(\theta) + V_{2c}(\theta).$$

We derive these components in the following. First, we have

$$\begin{aligned}V_1(\theta) &= \sum_{\{(r_1,i_1),(r_2,i_2)\}} \text{Var}\left(I\left\{\frac{X_{(r_1)i_1}+X_{(r_2)i_2}}{2} > \theta_0\right\}\right) \\ &= \frac{n^2}{2}\sum_{r_1=1}^{k}\sum_{r_2=1}^{k} \text{Var}\left(I\left\{\frac{X_{(r_1)1}+X_{(r_2)2}}{2} > \theta_0\right\}\right) \\ &+ \frac{n}{2}\sum_{r=1}^{k}\left[2\text{Var}(I\{X_{(r)} > \theta_0\}) - \text{Var}\left(I\left\{\frac{X_{(r)1}+X_{(r)2}}{2} > \theta_0\right\}\right)\right],\end{aligned}$$

$$
\begin{aligned}
V_{2a}(\theta) &= n(n-1)(n-2) \sum_{r,r_1,r_2=1}^{k} C([(r,1),(r_1,2)],[(r,1),(r_2,3)]) \\
&= n(n-1)(n-2)k^2 \sum_{r=1}^{k} Cov(I\{X_{(r),1}+X>2\theta_0\}, I\{X_{(r),1}+Y>2\theta_0\}) \\
&= n(n-1)(n-2)k^2\left[kP(X_1+X_2>2\theta_0, X_1+X_3>2\theta_0) - \sum_{r=1}^{k} P^2(X_1+X_2>2\theta_0)\right].
\end{aligned}
$$

To derive $V_{2b}(\theta)$, we consider separately two situations: $C([(r,i_1),(r_1,i_2)],[(r,i_1),(r_2,i_2)])$ and $C([(r,i_1),(r_1,i_1)],[(r,i_1),(r_2,i_2)])$. Denote the sum of the terms in the first situation by $V_{2b}^{(1)}(\theta)$ and that in the second situation by $V_{2b}^{(2)}(\theta)$. We have

$$
\begin{aligned}
V_{2b}^{(1)}(\theta) &= n(n-1)\sum_{r=1}^{k}\sum_{r_1\neq r_2} C([(r,1),(r_1,2)],[(r,1),(r_2,2)]) \\
&= n(n-1)\sum_{r=1}^{k}\sum_{r_1\neq r_2} C([(r,1),(r_1,2)],[(r,1),(r_2,3)]) \\
&= \frac{k^3 n(n-1)}{6(k+1)} - n(n-1)\sum_{r=1}^{k}\sum_{s=1}^{k}[P(X_{(r),1}+X_{(s),2}>2\theta_0, X_{(r),1}+X_{(s),3}>2\theta_0) \\
&\quad -P(X_{(r),1}+X_{(s),2}>2\theta_0)P(X_{(r),1}+X_{(s),2}>3\theta_0)].
\end{aligned}
$$

$$
\begin{aligned}
V_{2b}^{(2)}(\theta) &= 2n(n-1)\sum_{r=1}^{k}\sum_{r_1,r_2=1}^{k} C([(r,1),(r_1,1)],[(r,1),(r_2,2)]) \\
&= 2n(n-1)\sum_{r=1}^{k}\sum_{r_1,r_2=1}^{k} C([(r,1),(r_1,2)],[(r,1),(r_2,3)]) \\
&\quad +2n(n-1)\sum_{r=1}^{k}\sum_{r_2=1}^{k}[C([(r,1),(r,1)],[(r,1),(r_2,2)]) \\
&\quad -C([(r,1),(r,2)],[(r,1),(r_2,3)])] \\
&= \frac{2k^3 n(n-1)}{6(k+1)} + 2n(n-1)k\sum_{r=1}^{k}[P(X_{(r),1}>\theta_0, X_{(r),1}+X_2>2\theta_0) \\
&\quad -P(X_{(r),1}>\theta_0)P(X_{(r),1}+X_2>2\theta_0) \\
&\quad -P(X_{(r),1}+X_{(r),2}>2\theta_0, X_{(r),1}+X_3>2\theta_0) \\
&\quad +P(X_{(r),1}+X_{(r),2}>2\theta_0)P(X_{(r),1}+X_3>2\theta_0)].
\end{aligned}
$$

Finally, we derive that

$$
\begin{aligned}
V_{2c}(\theta) &= n\sum_{r=1}^{k}\sum_{r_1\neq r_2} C([(r,1),(r_1,1)],[(r,1),(r_2,1)]) \\
&= n\sum_{r=1}^{k}\sum_{r_1,r_2=1}^{k} C([(r,1),(r_1,2)],[(r,1),(r_2,3)]) \\
&\quad +2n\sum_{r=1}^{k}\sum_{s=1}^{k}[C([(r,1),(r,1)],[(r,1),(s,3)]) - C([(r,1),(r,2)],[(r,1),(s,3)])] \\
&\quad -2n\sum_{r=1}^{k}[C([(r,1),(r,1)],[(r,1),(r,3)]) - C([(r,1),(r,2)],[(r,1),(r,3)])] \\
&\quad -n\sum_{r,s=1}^{k} C([(r,1),(s,2)],[(r,1),(s,3)]).
\end{aligned}
$$

Under the null hypothesis, it can be derived that

$$
\begin{aligned}
V_1(\theta_0) &= \frac{n^2k^2}{4} - \frac{n^2}{2}\sum_{r,s=1}^{k}\psi_{rs} + \frac{nk}{2} - n\sum_{r=1}^{k}p_r^2(0) - \frac{n}{2}\sum_{r=1}^{k}(\psi_{rr}-\psi_{rr}^2) \\
&= n^3k^2\sum_{r=1}^{k}\frac{r(k-r+1)}{(k+1)^2(k+2)} - n(3n-2)\sum_{r,s=1}^{k}(\phi_{rs}-\psi_{rs}^2) \\
&= \frac{n^3k^3}{6(k+1)} - n(3n-2)\sum_{r,s=1}^{k}(\phi_{rs}-\psi_{rs}^2),
\end{aligned}
$$

where $p_r(0)$ is defined in (5.7) and

$$
\begin{aligned}
\phi_{rs} &= P(U_{(r)1}+U_{(s)2}>1, U_{(r)1}+U_{(s)3}>1) \qquad (5.15) \\
&= \frac{k!^3}{(r-1)!(k-r)!(s-1)!^2(k-s)!^2}\times \\
&\quad \int_{\substack{0<u_1<1\\ u_1+u_2>1\\ u_1+u_3>1}} u_1^{r-1}(u_2u_3)^{s-1}(1-u_1)^{k-r}\times \\
&\qquad [(1-u_2)(1-u_3)]^{k-s}du_1du_2du_3 \\
&= \frac{k!^3}{(r-1)!(k-r)!(s-1)!^2(k-s)!^2}\times \\
&\quad \int_{0<u_2,u_3<u_1<1} u_1^{r-1}[(1-u_2)(1-u_3)]^{s-1}\times \\
&\qquad (1-u_1)^{k-r}(u_2u_3)^{k-s}du_1du_2du_3
\end{aligned}
$$

$$
\begin{aligned}
&= \frac{k!^3}{(r-1)!(k-r)!(s-1)!^2(k-s)!^2} \times \\
&\int_{0<u_1,u_2,u_3<1} u_1^{r+1}[(1-u_1u_2)(1-u_1u_3)]^{s-1} \times \\
&\qquad (1-u_1)^{k-r}(u_1^2u_2u_3)^{k-s}du_1du_2du_3 \\
&= \sum_{t_1,t_2=0}^{s-1} \binom{2k+r-t_1-t_2-1}{r-1,k-t_1,k-t_2}\binom{k-r+t_1+t_2}{k-r,t_1,t_2}\Big/\binom{3k}{k,k,k},
\end{aligned}
$$

$$
\begin{aligned}
\psi_{rs} &= P(U_{(r)1} + U_{(s)2} > 1) \qquad (5.16)\\
&= \sum_{t=0}^{s-1}\binom{k-r+t}{t}\binom{k+r-t-1}{r-1}\Big/\binom{2k}{k},
\end{aligned}
$$

and that

$$
\begin{aligned}
V_{2a}(\theta_0) &= n(n-1)(n-2)\left[\frac{k}{3} - \sum_{r=1}^{k}\left(\frac{r}{k+1}\right)^2\right] \\
&= \frac{k^3 n(n-1)(n-2)}{6(k+1)}, \\
V_{2b}(\theta_0) &= \frac{k^3 n(n-1)}{2(k+1)} - n(n-1)\sum_{r=1}^{k}\sum_{s=1}^{k}[\phi_{r,s} - \psi_{r,s}^2] \\
&\quad +2n(n-1)\left[\frac{3k^2}{8} - \sum_{r,s=1}^{k}\phi_{r,s} - \frac{1}{k+1}\sum_{r=1}^{k}(rp_r - r\psi_{rr})\right], \\
V_{2c}(\theta_0) &= \frac{k^3 n}{6(k+1)} + 2n\left[\frac{3k(k-1)}{8} - \sum_{r\neq s=1}^{k}\phi_{r,s} + \frac{k-1}{k+1}\sum_{r=1}^{k}(rp_r - r\psi_{rr})\right] \\
&\quad -n\sum_{r,s=1}^{k}[\phi_{r,s} - \psi_{r,s}^2].
\end{aligned}
$$

Here we have used the fact that uniform distribution can be assumed and that $P(X_{(r)} + X_2 > 1) = EX_{(r)} = r/(k+1)$. Note that $V(\theta_0)$ is dominated by the term $\frac{n^3k^3}{6(k+1)}$.

By using similar arguments to those used for the central limit theorem of the one-sample U-statistic, we can easily prove the following theorem.

Theorem 5.5. *As $n \to \infty$, we have*

$$(W_{RSS}^+ - \mu(\theta))/\sqrt{V(\theta)} \to N(0,1).$$

In particular, under the null hypothesis,

$$[W_{RSS}^+ - \frac{nk(nk+1)}{4}]\Big/\sqrt{n^3k^3} \to N\left(0, \frac{1}{6(k+1)}\right).$$

From this theorem and (5.13), we immediately obtain the asymptotic relative efficiency of the RSS version of the Wilcoxon signed rank test with respect to its SRS counterpart as follows.

$$\mathrm{ARE}(W^+_{\mathrm{RSS}}, W^+_{\mathrm{SRS}}) = \frac{k+1}{2}.$$

It can also be argued that the null distribution of W^+_{RSS} is free of the underlying distribution. By the symmetry of F, we have that, under the null hypothesis,

$$\begin{aligned} W^+_{\mathrm{RSS}} &= \sum_{\{(r_1,i_1),(r_2,i_2)\}} I\{(X_{(r_1)i_1} - \theta_0 > -X_{(r_2)i_2} + \theta_0\} \\ &= \sum_{\{(r_1,i_1),(r_2,i_2)\}} I\{F(X_{(r_1)i_1} - \theta_0) > 1 - F(X_{(r_2),i_2} - \theta_0)\} \\ &\stackrel{\mathcal{D}}{=} \sum_{\{(r_1,i_1),(r_2,i_2)\}} I\{U_{(r_1)i_1} + U_{(r_2)i_2} > 1\}, \end{aligned}$$

where $\{U_{(r)i} : r = 1, \cdots, k; i = 1, \cdots, n\}$ is a RSS sample from the uniform distribution $\mathcal{U}(0,1)$. Therefore, the distribution of W^+_{RSS} does not depend on F under the null hypothesis.

5.3.2 Decision rules of the Wilcoxon signed rank test

The rejection region of the Wilcoxon signed rank test for testing $H_0 : \theta = \theta_0$ takes the form:

$$\begin{aligned} \left| W^+_{RSS} - \frac{nk(nk+1)}{4} \right| &> c_n, \\ W^+_{RSS} - \frac{nk(nk+1)}{4} &> c_n, \\ \text{or } W^+_{RSS} - \frac{nk(nk+1)}{4} &< c_n, \end{aligned}$$

where c_n is a generic critical value determined by the significance level, according as H_1 is

$$\begin{aligned} \theta &\neq \theta_0, \\ \theta &> \theta_0, \\ \text{or } \theta &< \theta_0. \end{aligned}$$

For example, for the two-sided test $H_0 : \theta = \theta_0$ v.s. $H_1 : \theta \neq \theta_0$, the critical value c_n is determined as follows. When n is large, by the asymptotic distribution of W^+_{RSS},

$$c_n = z_{\alpha/2}\sqrt{(nk)^3/6(k+1)}.$$

If n is moderate, the asymptotic variance of W^+_{RSS} in the normal approximation is replaced by its actual variance, then

$$c_n = [z_{\alpha/2}\sqrt{V(\theta_0)} - 0.5].$$

If n is small,the exact null distribution of W^+_{RSS} is needed. However its computation is complicated. As an illustration, consider the case that $n = 1$ and $k = 2$. In this case, there are only two observations in the RSS sample, (say, $U_{(1)}$ and $U_{(2)}$), and W^+_{RSS} may take 4 values, $0, 1, 2, 3$. The null distribution of W^+_{RSS} is as follows.

$$\begin{aligned}
P(W^+_{\text{RSS}} = 0) &= P(U_{(1)} < 1/2, U_{(2)} < 1/2) = \frac{3}{16} \\
P(W^+_{\text{RSS}} = 1) &= P(U_{(2)} > 1/2, U_{(1)} + U_{(2)} < 1) \\
&\quad + P(U_{(1)} > 1/2, U_{(1)} + U_{(2)} < 1) = \frac{5}{16} \\
P(W^+_{\text{RSS}} = 2) &= P(U_{(1)} < 1/2, U_{(1)} + U_{(2)} > 1) \\
&\quad + P(U_{(2)} < 1/2, U_{(1)} + U_{(2)} > 1) = \frac{5}{16} \\
P(W^+_{\text{RSS}} = 3) &= P(U_{(1)} > 1/2, U_{(2)} > 1/2) = \frac{3}{16}.
\end{aligned}$$

In general, a computationally intensive computer program needs to be worked out for the computation of the null distribution. The critical value c_n is then taken as the smallest integer such that

$$P(W^+_{\text{RSS}} \le \frac{nk(nk+1)}{4} - c_n) \le \alpha/2.$$

5.3.3 Estimation of θ

In general, the population median can be estimated by the ranked set sample median as discussed in Section 2.6. However, a more efficient estimate using the ranked set sample can be obtained by taking into account the symmetry of the underlying distribution.

Let X_1, X_2 be independent and identically distributed as X. If X follows a symmetric distribution with median θ, then $(X_1+X_2)/2$ follows a distribution with the same median. If we define an empirical distribution on the averages $\left\{\frac{X_{(r_1)i_1}+X_{(r_2)i_2}}{2}\right\}$ then, by following the derivation of EW^+_{RSS}, we can see that the empirical distribution provides an asymptotically unbiased estimate of the distribution of $(X_1 + X_2)/2$. Hence, heuristically, the median of the averages will provide a reasonable estimate of θ. Denote this median by $\hat{\theta}_W$. In what follows, we show that, indeed, $\hat{\theta}_W$ is a better estimator than the ranked set sample median.

Let $\theta_N = \theta + x/\sqrt{N}$ and

$$W_{\text{RSS}}^{+}(\theta_N) = \sum_{\{(r_1,i_1),(r_2,i_2)\}} I\{(X_{(r_1)i_1} + X_{(r_2)i_2})/2 > \theta_N\}).$$

Then we have,

$$\sqrt{N}(\hat{\theta}_W - \theta) \leq x \Leftrightarrow W_{\text{RSS}}^{+}(\theta_N) \leq \frac{N(N+1)}{4}.$$

Thus, it follows from the central limit theorem on $W_{\text{RSS}}^{+}(\theta_N)$ that

$$\begin{aligned}
&P(\sqrt{N}(\hat{\theta}_W - \theta) \leq x) \\
&= P(W_{\text{RSS}}^{+}(\theta_N) \leq \frac{N(N+1)}{4}) \\
&= P\left(\frac{W_{\text{RSS}}^{+}(\theta_N) - EW_{\text{RSS}}^{+}(\theta_N)}{\sqrt{\text{Var}(W_{\text{RSS}}^{+}(\theta_N))}} \leq \frac{\frac{N(N+1)}{4} - EW_{\text{RSS}}^{+}(\theta_N)}{\sqrt{\text{Var}(W_{\text{RSS}}^{+}(\theta_N))}}\right) \\
&\to \Phi(x f_2(0)\sqrt{6(k+1)}).
\end{aligned}$$

That is,

$$\sqrt{N}(\hat{\theta}_W - \theta) \to N\left(0, \frac{1}{6(1+k)f_2^2(0)}\right).$$

We can now make a comparison between $\hat{\theta}_W$ and the ranked set sample median. Recall from Section 2.6 that the ARE of the ranked set sample median with respect to the simple random sample median is given by

$$\frac{1}{4}\{\frac{1}{k}\sum_{r=1}^{k} B(r, k-r+1, 1/2)[1 - B(r, k-r+1, 1/2)]\}^{-1}.$$

The ARE of $\hat{\theta}_W$ with respect to the simple random sample median is given by

$$\frac{1}{4f^2(0)} \Big/ \frac{1}{6(1+k)f_2^2(0)} = \frac{3(k+1)}{2}\left[\frac{f_2(0)}{f(0)}\right]^2.$$

Thus, the ARE of $\hat{\theta}_W$ with respect to the ranked set sample median is given by

$$6(k+1)\left[\frac{f_2(0)}{f(0)}\right]^2 \frac{1}{k}\sum_{r=1}^{k} B(r, k-r+1, 1/2)[1 - B(r, k-r+1, 1/2)].$$

If the underlying distribution is Normal, we have that $\left[\frac{f_2(0)}{f(0)}\right]^2 = 1/2$ and that, for $k = 3, 4, 5$, the AREs are $1.875, 2.049, 2.217$. The estimator $\hat{\theta}_W$ is much more efficient than the ranked set sample median.

Denote the order statistics of the averages by $W_{(1)} \leq \cdots \leq W_{(\frac{N(N+1)}{2})}$. Then a $100(1-\alpha)\%$ confidence interval of θ can be constructed as

$$[\hat{\theta}_L, \ \hat{\theta}_U],$$

where

$$\hat{\theta}_L = W_{(\frac{N(N+1)}{4} - c_n)} \text{ and } \hat{\theta}_U = W_{(\frac{N(N+1)}{4} + c_n)},$$

if $\frac{N(N+1)}{4}$ is an integer,

$$\hat{\theta}_L = W_{(\frac{N(N+1)}{4} + 0.5 - c_n)} \text{ and } \hat{\theta}_U = W_{(\frac{N(N+1)}{4} + 0.5 + c_n)},$$

otherwise, where c_n is determined in exactly the same way as in the two-sided test.

5.3.4 Effect of imperfect ranking on RSS Wilcoxon signed rank test

The remark on imperfect ranking made for the MWW test applies to the Wilcoxon signed rank test as well. In general, an RSS version of the Wilcoxon signed rank test is more efficient than its SRS counterpart no matter whether ranking is perfect or not. However, when ranking is imperfect, the efficiency is less than when ranking is perfect. There is a particular feature with the Wilcoxon signed rank test. For the sign test and the MWW test discussed in the previous sections, though the ranking errors affect the variances of the test statistics, the means of the test statistics are not affected by the ranking errors at all. But both the variance and the mean of the Wilcoxon signed rank test statistic involve the ranking error probabilities. However, it can be shown that the asymptotic mean of the test statistic is still free of the ranking error probabilities. Furthermore, as in the perfect ranking case, the asymptotic normality of the Wilcoxon signed rank test statistic with imperfect ranking can be obtained. Again, the problem is that the asymptotic distribution under the null hypothesis can not be completely determined without knowing the ranking error probabilities. Again, to make an inference, we have to resort to the bootstrap method mentioned at the end of the last section.

5.4 Optimal design for distribution-free tests

We consider the optimal design of unbalanced RSS for distribution-free tests in this section. The optimal design for the sign test is discussed in Section 5.4.1. The optimal design for the Wilcoxon signed rank test is discussed in Section 5.4.2.

5.4.1 Optimal design for sign test

Consider the general unbalanced RSS scheme discussed in Section 4.1. Recall that, under the general scheme, among n ranked samples of size k, the rth ranked order statistic is measured for n_r of the ranked samples for $r = 1, \dots, k$. Let $X_{(r)i}, r = 1, \dots, k, i = 1, \dots, n_r$, denote the measured ranked order statistics. Define

$$S_g^+ = \sum_{r=1}^{k} \sum_{i=1}^{n_r} I(X_{(r)i} > \theta_0).$$

Then

$$ES_g^+ = \sum_{r=1}^{k} n_r \bar{F}_{(r))}(\theta_0 - \theta)$$

and

$$\text{Var}(S_g^+) = \sum_{r=1}^{k} n_r \bar{F}_{(r)}(\theta_0 - \theta) F_{(r))}(\theta_0 - \theta).$$

We have

$$\frac{d}{d\theta} E(S_g^+)\bigg|_{\theta_0} = n \sum_{r=1}^{k} q_r f_r(0) = n \sum_{r=1}^{k} q_r \frac{k! 2^{-k+1}}{(r-1)!(k-r)!} f(0),$$

where $q_r = n_r/n$.

Under null hypothesis, we have

$$\text{Var}(S_g^+) = n \sum_{r=1}^{k} q_r p_r (1 - p_r)$$

where p_r is defined in (5.7).

Therefore, we have the following theorem.

Theorem 5.6. *The efficacy of S_g^+ is given by*

$$Eff(S_g^+) = \frac{[\sum_{r=1}^{k} q_r \frac{k! 2^{-k+1}}{(r-1)!(k-r)!} f(0)]^2}{\sum_{r=1}^{k} q_r p_r (1 - p_r)}.$$

(i) Optimal Design of Maximum Efficacy. By Lemma 4.7 in Section 4.6, the maximum efficacy attains when at most two components of $\boldsymbol{q} = (q_1, \dots, q_k)^T$, say q_r and q_s, are nonzero. Suppose $r < s$. Noting that

$$\frac{k! 2^{-k+1}}{(s-1)!(k-s)!} f(0) = \frac{k! 2^{-k+1}}{((k+1-s)-1)!(k-(k+1-s))!} f(0)$$

and

$$p_s(1 - p_s) = p_{k+1-s}(1 - p_{k+1-s}),$$

we may further assume that $r < s \le (k+1)/2$ and that $q_r = \lambda = 1 - q_s$.

Now, the determination of the optimal design reduces to the maximization of

$$Q(\lambda) = \frac{[\lambda(a_1 - a_2) + a_2]^2}{\lambda(b_1 - b_2) + b_2}$$

with respect to $\lambda \in [0, 1]$, where $a_1 = \frac{1}{(s-1)!(k-s)!}$, $a_2 = \frac{1}{(r-1)!(k-r)!}$, $b_1 = p_s(1 - p_s)$ and $b_2 = p_r(1 - p_r)$.

The solution to $\frac{d}{d\lambda}Q(\lambda) = 0$ is unique and equals

$$\lambda = \frac{a_2 b_1 - 2a_1 b_2 + a_2 b_2}{(a_1 - a_2)(b_1 - b_2)}.$$

Note that $Q(\lambda) \to +\infty$ as $\lambda \to \infty$ or $-b_2/(b_1 - b_2)$. Thus, the above solution is a minimizer of $Q(\lambda)$ and hence the maximum of $Q(\lambda)$ can only be reached at $\lambda = 0$ or 1. We have proved the following result.

Theorem 5.7. *The maximal efficacy of S_g^+ is reached when only one component of $\boldsymbol{q}$ is nonzero, that is,*

$$\sup_{\boldsymbol{q}} Eff(S_g^+) = f^2(0) \max_{1 \le r \le (k+1)/2} \frac{\left(\frac{m! 2^{-k+1}}{(r-1)!(k-r)!}\right)^2}{p_r(1 - p_r)}.$$

Computed values of $[\frac{k! 2^{-k+1}}{(j-1)!(k-j)!}]/\sqrt{p_j(1 - p_j)}$ show that the maximum occurs at $r = [(k+1)/2]$. That is, the maximum efficacy of S_g^+ attains when only the $[(k+1)/2]$-th order statistic is measured for all the ranked sets. However, the theoretical proof seems very involved. The values of $[\frac{k! 2^{-k+1}}{(j-1)!(k-j)!}]/\sqrt{p_j(1 - p_j)}$ are presented in Table 5.7 for $k \le 16$.

(ii) ARE of the Optimal Sign Test to Balanced RSS Sign Test. It is easy to show that

$$ARE(S^+_{Opt}, S^+_{RSS}) = \frac{\delta_0^2 \left(\frac{k! 2^{-k+1}}{(i_k - 1)!(k - i_k)!}\right)^2}{4 p_{i_k}(1 - p_{i_k})},$$

where $i_k = [(k+1)/2]$.

The ARE values for $k \le 16$ are given in Table 5.8.

5.4.2 Optimal design for Wilcoxon signed rank tests

We now discuss the statistical properties of the Wilcoxon signed rank tests based on unbalanced RSS for seeking higher efficiency of the sampling scheme. In this subsection, we introduce some results obtained by Öztürk and Wolfe [120].

(i) Quasi-balanced RSS. For a positive integer $t \le k$, let $D = \{d_1, \cdots, d_t\}$ be a subset of $\{1, \cdots, k\}$ of size t. The entries of D will be the ranks of the observations to be quantified in each cycle. For example, if $D = \{1, 2, 5\}$ and

Table 5.7. One order Efficacy of $(S_g^+)^2(/f^2(0))$ $k = 2, \cdots, 16$ (with value of $k+1-j$ = value of j)

$k \backslash j$	1	2	3	4	5	6	7	8
2	5.333							
3	5.143	9.000						
4	4.267	10.473						
5	3.226	10.256	14.063					
6	2.286	9.023	15.584					
7	1.543	7.350	15.361	19.141				
8	1.004	5.643	13.933	20.687				
9	0.634	4.131	11.850	20.460	24.225			
10	0.391	2.908	9.563	18.911	25.787			
11	0.236	1.981	7.384	16.543	25.558	29.312		
12	0.141	1.313	5.491	13.813	23.925	30.884		
13	0.083	0.850	3.951	11.077	21.347	30.655	34.401	
14	0.048	0.540	2.764	8.574	18.271	28.961	35.980	
15	0.027	0.337	1.887	6.431	15.071	26.222	35.750	39.491
16	0.016	0.207	1.260	4.691	12.028	22.868	34.010	41.075

Table 5.8. $ARE(S_{Opt}^+, S_{RSS}^+)$ $k = 2, \cdots, 16$

k	2	3	4	5	6	7	8	9
ARE	1.0	1.406	1.432	1.730	1.758	2.005	2.031	2.247
k	10	11	12	13	14	15	16	
ARE	2.272	2.465	2.489	2.666	2.689	2.853	2.874	

$k = 5$, we would quantify two extreme and one rank 2 observations in each cycle. We shall call D the *Design*. For a given D, we collect data as follows: we draw kt units from an infinite population with symmetric and absolutely continuous distribution $F(\cdot)$, and partition them into t sets of sizes k. Each set is judgment ranked without actually measuring the units. In the first set, we measure the observation of rank d_1. In the second set, we measure the observation of rank d_2. Continue the procedure until the first cycle $X_{(d_1)1}, \cdots, X_{(d_t)1}$ is complete. Then the procedure is repeated m times. We refer to this sampling scheme as a *quasi-balanced RSS*. We shall consider optimal designs based on quasi-balanced RSS.

(ii) Signed rank test with selective designs. In the following discussion, we assume $\theta_0 = 0$ without loss of generality. With quasi-balanced RSS, our signed rank test can be rewritten as

$$W_D^+ = \sum_{i=1}^{t}\sum_{j=1}^{m} I(X_{(d_i)j} > 0)R^+(|X_{(d_i)j}|). \tag{5.17}$$

Lemma 5.8. *Under H_0: the population distribution is continuous and symmetric about 0, W_D^+ is distribution free.*

The conclusion of the lemma is an easy consequence of the following fact

$$\sum_{i=1}^{t}\sum_{j=1}^{m} I(X_{(d_i)j} > 0)R^+(|X_{(d_i)j}|) = \sum_{i=1}^{t}\sum_{j=1}^{m} I(Y_{(d_i)j} > \frac{1}{2})R^+(|Y_{(d_i)j} - \frac{1}{2}|)$$

where $Y_{(d_i)j} = F(X_{(d_i)j})$ and F is the population distribution function. The right hand side of the above equation is the signed rank test with quasi-balanced RSS from the uniform distribution.

Lemma 5.9. *Under H_0, if the design is symmetric, i.e., $d_i \in D$ implies $d_{k+1-i} \in D$, then*

$$EW_D^+ = \frac{mt(mt+1)}{4}. \tag{5.18}$$

Proof. By symmetry, we have $X_{(d_i)j} \stackrel{\mathcal{D}}{=} -X_{(k+1-d_i)j}$. Thus,

$$\begin{aligned}
EW_D^+ &= E\sum_{i=1}^{t}\sum_{j=1}^{m} I(-X_{(k+1-d_i)} > 0)R^+(|X_{(k+1-d_i)}|) \\
&= E\sum_{i=1}^{t}\sum_{j=1}^{m} (1 - I(X_{(k+1-d_i)} > 0))R^+(|X_{(k+1-d_i)}|) \\
&= E\sum_{i=1}^{t}\sum_{j=1}^{m} (1 - I(X_{(d_i)} > 0))R^+(|X_{(d_i)}|) \\
&= \frac{1}{2}E\sum_{i=1}^{t}\sum_{j=1}^{m} R^+(|X_{(d_i)}|) \\
&= \frac{mt(mt+1)}{4}.
\end{aligned}$$

In general, to consider the limiting distribution of W_D^+, we decompose W_D^+ as

$$W_D^+ = mtV_D + mT_D + \binom{m}{2}t^2U_D,$$

where

$$V_D = \frac{1}{mt}\sum_{i=1}^{t}\sum_{j=1}^{m} I(X_{(d_i)j} > 0),$$

$$T_D = \frac{1}{m} \sum_{j=1}^{m} \sum_{1 \le i < r \le t} I(X_{(d_i)j} + X_{(d_r)j} > 0)$$

and

$$U_D = \frac{2}{m(m-1)t^2} \sum_{1 \le j < s \le m} \sum_{i,r=1}^{t} I(X_{(d_i)j} + X_{(d_r)s} > 0).$$

From this decomposition, we get the following lemmas.

Lemma 5.10. *Under the null hypothesis, the means of the three components in the decomposition of W_D^+ are given below.*

$$v_D = EV_D = \frac{1}{t} \sum_{i=1}^{t} p_{d_i}(0),$$

$$t_D = ET_D = \sum_{1 \le i < r \le t} P(X_{(d_i)} + X_{(d_r)} > 0) = \sum_{1 \le i < r \le t} \psi_{d_i,d_r}$$

and

$$u_D = EU_D = \frac{1}{t^2} \sum_{i,r=1}^{t} P(X_{(d_i)1} + X_{(d_r)2} > 0) = \frac{1}{t^2} \sum_{i,r=1}^{t} \psi_{d_i,d_r}, \tag{5.19}$$

where $p_r(0)$ and ψ_{rs} are defined in (5.7) and (5.16), respectively.

Lemma 5.11. *The variances of the three components are given below:*

$$\sigma^2(V_D) = \frac{1}{mt^2} \sum_{i=1}^{t} [p_{d_i}(0) - p_{d_i}^2(0)],$$

$$\sigma^2(T_D) = \frac{1}{m} \sum_{1 \le i < r \le t} [\psi_{d_i,d_j} - \psi_{d_i,d_j}^2] + \frac{1}{m} \sum_{i \ne r \ne s \le t} [\psi_{d_i:d_r,d_s} - \psi_{d_i,d_r}\psi_{d_i,d_s}]$$

$$\sigma^2(U_D) = \sigma_{1D}^2 + \sigma_{2D}^2,$$

where

$$\sigma_{1D}^2 = \frac{2}{m(m-1)t^4} \sum_{i,r=1}^{t} [\psi_{d_i,d_r} - \psi_{d_i,d_r}^2] + \frac{4}{m(m-1)t^4} \sum_{i=1}^{t} \sum_{r_1 \ne r_2} [\psi_{d_i:d_{r_1},d_{r_2}} - \psi_{d_i,d_{r_1}}\psi_{d_i,d_{r_2}}],$$

$$\sigma_{2D}^2 = \frac{4(m-2)}{m(m-1)t^4} \sum_{i,r_1,r_2=1}^{t} [[\psi_{d_i:d_{r_1},d_{r_2}} - \psi_{d_i,d_{r_1}}\psi_{d_i,d_{r_2}}]$$

and

$$\begin{aligned}
\psi_{i:r,s} &= P(U_{(i)1} + U_{(r)2} > 1, U_{(i)1} + U_{(s)3} > 1) \\
&= \frac{(k!)^3}{(i-1)!(r-1)!(s-1)!(k-i)!(k-r)!(k-s)!} \\
&\int_{\substack{x+y>1 \\ x+z>1}} x^{i-1}(1-x)^{k-i}y^{r-1}(1-y)^{k-r}z^{s-1}(1-z)^{k-s}dxdydz \\
&= \sum_{t_1}^{r-1}\sum_{t_2=0}^{s-1} \binom{2k-t_1-t_2+i-1}{k-t_1, k-t_2, i-1}\binom{k+t_1+t_2-i}{k-i, t_1, t_2} \Big/ \binom{3k}{k,k,k}
\end{aligned}$$

Using the central limit theorem for one sample U-statistics, we have

Theorem 5.12. *Under the null hypothesis,*

$$\sigma_{2D}^{-1}(U_D - u_D) \Rightarrow N(0,1),$$

which implies that

$$\sqrt{\frac{2}{m(m-1)t^2}}\sigma_{2D}^{-1}(W_D^+ - EW_D^+) \Rightarrow N(0,1).$$

(iii) Pitman's ARE. Notice that the components V_D and T_D play no role in the limiting distribution of W_D^+. Thus, the Pitman's efficacy of W_D^+ is determined by that of U_D.

Rewrite u_D as $u_D(\theta)|_{\theta=0}$, then one can easily calculate

$$\begin{aligned}
u_D'(0) &= \frac{-2}{t^2}\sum_{i,r=1}^{t} f_{d_i} * f_{d_r}(0) \\
&= \frac{-2k^2}{t^2}\sum_{i,r=1}^{t}\binom{k-1}{d_i-1}\binom{k-1}{d_r-1}K_F(d_i,d_r),
\end{aligned}$$

where

$$K_F(i,r) = \int F^{k-i+r-1}(x)(1-F(x))^{k+i-r-1}f^2(x)dx.$$

We obtain the following theorem.

Theorem 5.13. *For a fixed design D, the Pitman's efficacy of W_D^+ is given by*

Table 5.9. Pitman ARE with one rank design and normal distribution

d_1	k	ARE	d_1	k	ARE
1	2	.877948	2	5	.836178
2	2	.877947	3	5	1.167743
1	3	.620418	4	5	.836176
2	3	1.120298	5	5	.236668
3	3	.620417	1	6	.134132
1	4	.396471	2	6	.615537
2	4	1.038297	3	6	1.106043
3	4	1.038296	4	6	1.106042
4	4	.396437	5	6	.615533
1	5	.236676	6	6	.133641

Table 5.10. Pitman ARE with two rank design and normal distribution

d_1	d_2	k	ARE	d_1	d_2	k	ARE
1	2	2	1.000735	2(3)	3(4)	5	1.059424
1 (2)	2(3)	3	.945163	2	4	5	1.093328
1	3	3	.944741	1(5)	2(6)	6	.379629
1 (3)	2(4)	4	.752535	1(4)	3(6)	6	.680258
1 (2)	3(4)	4	.955983	1(3)	4(6)	6	.865053
1	4	4	.864375	1(2)	5(6)	6	.873552
2	3	4	1.118016	1	6	6	.715697
1(4)	2(5)	5	.550714	2(4)	3(5)	6	.894450
1(3)	3(5)	5	.831377	2(3)	4(5)	6	1.060285
1(2)	4(5)	5	.935240	2	5	6	1.062956
1	5	5	.782184	3	4	6	1.164594

$$Eff(W_D^+) = \frac{k^4\left[\sum_{i,r=1}^{t}\binom{k-1}{d_i-1}\binom{k-1}{d_r-1}K_F(d_i,d_r)\right]}{\sum_{i,r_1,r_2=1}^{t}[\psi_{d_i:d_{r_1},d_{r_2}} - \psi_{d_i,d_{r_1}}\psi_{d_i,d_{r_2}}]}.$$

The Pitman's ARE of W_D^+ to the W_{RSS}^+ is

$$ARE(W_D^+, W_{RSS}^+) = \frac{kEff(W_D^+)}{6tk(k+1)\{\int f^2(x)dx\}^2}.$$

PROOF: As for the second conclusion, notice that in Section 5.3, $Eff(W_{RSS}^+) = 6k(k+1)\{\int f^2(x)dx\}^2$. Then, the expression of $ARE(W_D^+, W_{RSS}^+)$ follows by noticing that W_D^+ uses mt observations whereas W_{RSS}^+ uses mk observations.

(iv) Numerical Comparison. Öztürk and Wolfe [120] computed the values of $ARE(W_D^+, W_{RSS}^+)$ for $k \leq 6$ and $t = 1, 2$ with standard normal density f. However, in their calculation, they used an additional assumption that

Table 5.11. Pitman ARE with two rank design and beta(0.6,0.6) distribution

d_1	d_2	k	ARE	d_1	d_2	k	ARE
1	2	2	.999947	2(3)	3(4)	5	.156518
1 (2)	2(3)	3	.220447	2	4	5	.241919
1	3	3	3.024371	1(5)	2(6)	6	.050342
1 (3)	2(4)	4	.125504	1(4)	3(6)	6	.099387
1 (2)	3(4)	4	.324800	1(3)	4(6)	6	.198232
1	4	4	7.342311	1(2)	5(6)	6	.912457
2	3	4	.192809	1	6	6	21.592780
1(4)	2(5)	5	.079280	2(4)	3(5)	6	.120779
1(3)	3(5)	5	.153678	2(3)	4(5)	6	.179238
1(2)	4(5)	5	.551663	2	5	6	.343963
1	5	5	13.618680	3	4	6	.158169

Table 5.12. Pitman ARE with two rank design and the mixed normal distribution:0.5$[N(-1,1)+N(1,1)]$.

d_1	d_2	k	ARE	d_1	d_2	k	ARE
1	2	2	1.000000	2(3)	3(4)	5	.938446
1 (2)	2(3)	3	.885439	2	4	5	1.031798
1	3	3	1.023459	1(5)	2(6)	6	.329102
1 (3)	2(4)	4	.679152	1(4)	3(6)	6	.600109
1 (2)	3(4)	4	.933213	1(3)	4(6)	6	.814195
1	4	4	1.025248	1(2)	5(6)	6	.980812
2	3	4	1.014340	1	6	6	.954227
1(4)	2(5)	5	.485396	2(4)	3(5)	6	.778554
1(3)	3(5)	5	.760533	2(3)	4(5)	6	.963284
1(2)	4(5)	5	.975707	2	5	6	1.067078
1	5	5	.994148	3	4	6	1.015016

the selected designs are symmetric when they derive the variance of U_D. By dropping this additional assumption, these values are recomputed and given in Tables 5.9 and 5.10. In the tables, the numbers in parentheses refer the ranks of an equivalent alternative design.

Based on their numerical results, Öztürk and Wolfe [120] claimed that an optimal sampling protocol "quantifies only the middle observation when the set size is odd and the two middle observations when the set size is even". However, this phenomenon might be only associated with the unimodality of the underlying distribution. It is not a universal phenomenon. We computed the values of $ARE(W_D^+, W_{RSS}^+)$ for a beta distribution, *Beta(0.6,0.6)*, and a mixed normal distribution, $0.5[N(-1,1)+N(1,1)]$, and found that the optimal design should quantify the observations whose corresponding quantiles are close to the modes. The values of $ARE(W_D^+, W_{RSS}^+)$ for the beta distribution and the mixed normal distribution are given in Tables 5.11 and 5.12 respectively. Because the modes of *Beta(0.6,0.6)* are at two end points, it is seen from Table 5.11 that the optimal design is to quantify the two extreme observations. For the normal mixture, for $t = 2$ and $k = 6$, the optimal design is $D = \{2, 5\}$. This is because the modes of the normal mixture are located in the lower and upper tails.

5.5 Bibliographic notes

The first work on distribution-free tests with RSS was due to Stokes and Sager [162] who studied the empirical distribution function defined on an RSS sample and investigated the improvement of the Kolmogorov-Smirnov test based on the RSS over the SRS version. In the sequel, Bohn and Wolfe [27] [28] investigated the RSS Mann-Whitney-Wilcoxon two-sample rank test. The RSS sign test was investigated by Hettmansperger [60], Koti and Babu [79], Öztürk and Wolfe [117] and Barabesi [8]. The statistical computation related to the RSS sign test was considered by Barabesi [7]. Bohn [26], Kim and Kim [76] and Öztürk and Wolfe [120] studied the RSS Wilcoxon signed rank test. The material in this chapter is mainly based on the work of the above authors, but has been re-derived and presented in a much simpler form.

6

Ranked Set Sampling with Concomitant Variables

There are abundant practical examples where a sampling unit is associated with several variables, one of which is expensive to measure but the others can be obtained easily and cheaply. The "expensive" variable is either the variable of main interest or the response variable in the setting of a regression model. A few such examples were given in Chapter 1. For the sake of convenience, we refer to the "expensive" variable as the response variable in all cases and to the others as concomitant variables. In this chapter, we are concerned with the estimation of certain characteristics of the response variable and the estimation of the regression model. We deal with how concomitant variables can be used in RSS and the related regression analysis. The use of a single concomitant variable in RSS was briefly mentioned in Chapter 2. It was shown there the ranking mechanism using a single concomitant variable is consistent. In the current chapter, we concentrate on consistent ranking mechanisms using multiple concomitant variables. Two such ranking mechanisms are developed in this chapter: a multi-layer ranking mechanism and an adaptive ranking mechanism. The RSS with the multi-layer ranking mechanism, which is referred to as the multi-layer RSS, is discussed in Section 6.1. The multi-layer ranking mechanism is conceptually equivalent to a stratification of the space of the concomitant variables. The multi-layer RSS is particularly useful for the estimation of regression coefficients. The features of the multi-layer RSS are investigated through simulation studies. The RSS with the adaptive ranking mechanism, which is referred to as the adaptive RSS, is discussed in Section 6.2. In the adaptive RSS, the conditional expectation of the response variable given the concomitant variables is used as ranking criterion in an adaptive way. The conditional expectation is continually estimated and updated using the data already obtained and then used as ranking criterion for further sampling. The adaptive ranking mechanism is the best, at least asymptotically, if the major concern is the estimation of certain characteristics of the response variable such as its mean and quantiles. In Section 6.3, the estimation of the regression model and regression-estimates of the mean of the response variable in the context of RSS are discussed. It is argued that, for the estimation of the mean of the response variable, the RSS regression-estimate is better than the RSS sample mean as long as the response variable and the concomitant variables are moderately correlated. It is shown that, for the estimation of the regression coefficients, balanced RSS and SRS are asymptotically equivalent in terms of efficiency. For more efficient estimates of the regression coefficients, unbalanced RSS is in order. Section 6.4 takes up the design of unbalanced RSS for the regression analysis under the consideration of A-optimality, D-optimality and IMSE-optimality. A general method is discussed and details are provided for polynomial regression models. Some technical details are given in Section 6.5.

6.1 Multi-layer ranked set sampling

We develop in this section a sampling scheme using multiple concomitant variables which will be referred to as a multi-layer ranked set sampling (MRSS). To distinguish, an RSS using a single concomitant variable as the ranking criterion is to be called a marginal RSS. The motivation and definition of the MRSS is given in Section 6.1.1. The consistency of the MRSS is discussed in Section 6.1.2. The MRSS is compared with marginal RSS by simulation studies in Section 6.1.3. Issues concerning the choice of concomitant variables for the MRSS are considered in Section 6.1.4.

6.1.1 Motivation and definition

As a motivation, let us first consider RSS with a single concomitant variable. Let Y and X denote, respectively, the response variable and the concomitant variable. RSS using X as the ranking criterion goes as follows. Each time, a set of k units — a simple random sample of size k — is drawn from the population and the value of X is measured for all the units in the set, then the units are ranked according to the order of their X values. In the first ranked set, the unit with rank 1 is taken for the measurement of Y, in the second ranked set, the unit with rank 2 is taken for the measurement of Y, and so on, until every rank from 1 to k has an associated unit being taken for the measurement of Y. Then the cycle repeats. This procedure post-stratifies the sampling units into strata such that within each stratum the X-features of the units are more homogeneous. When Y and X are correlated, the Y-features of the units are also more homogeneous within each stratum. Therefore, the Y-features can be estimated more accurately.

Let us now turn to the case of multiple concomitant variables. We can represent the population by the sample space of the multiple concomitant variables. In RSS, to rank the sampling units with respect to a single concomitant variable is like partitioning the space into slices along one axis of the space. If every concomitant variable is reasonably associated with the response variable, the features of Y within each slice are still subject to the variation due to other concomitant variables and hence will not be as homogeneous as desired. An obvious alternative is then to partition the space into cubes along all axes. Then we can expect more homogeneity within a cube than within a slice. This leads us to the MRSS scheme to be described in the following.

Without loss of generality, consider two concomitant variables for the sake of convenience. Let $X^{\{1\}}$ and $X^{\{2\}}$ denote the concomitant variables. Let k, l be positive integers. A two-layer RSS procedure goes as follows. First, kl^2 independent sets, each of size k, are drawn from the population. The units in each of these sets are ranked according to $X^{\{1\}}$. Then, for l^2 ranked sets, the units with $X^{\{1\}}$-rank 1 are selected, for another l^2 ranked sets, the units with $X^{\{1\}}$-rank 2 are selected, and so on. Let the values of $(Y, X^{\{1\}}, X^{\{2\}})$ of these selected units be denoted by

$$\begin{array}{ccc} (Y_{[1]1}, X^{\{1\}}_{[1]1}, X^{\{2\}}_{[1]1}) & \cdots & (Y_{[1]l^2}, X^{\{1\}}_{[1]l^2}, X^{\{2\}}_{[1]l^2}) \\ \cdots & \cdots & \cdots \\ (Y_{[k]1}, X^{\{1\}}_{[k]1}, X^{\{1\}}_{[k]1}) & \cdots & (Y_{[k]l^2}, X^{\{1\}}_{[k]l^2}, X^{\{2\}}_{[k]l^2}) \end{array} \tag{6.1}$$

where the values of $(X^{\{1\}}_{[r]i}, (X^{\{2\}}_{[r]i})$ are measured and $Y_{[r]i}$ are latent. This completes the first layer of the procedure. In the second layer, the units represented in each row of (6.1) are divided — randomly or systematically — into l subsets, each of size l. The units in each of these subsets are ranked according to $X^{\{2\}}$. Then, for the first ranked subset, the unit with $X^{\{2\}}$-rank 1 are selected and its value on Y is measured, for the second ranked subset, the unit with $X^{\{2\}}$-rank 2 is selected and its value on Y is measured, and so on. This completes one cycle of the procedure. Repeating the cycle m times then yields the data set

$$\{Y_{[r][s]j} : r = 1, \ldots, k; s = 1, \ldots, l; j = 1, \ldots, m\}$$

where $Y_{[r][s]j}$ is the measurement of Y in the jth cycle on the unit with $X^{\{1\}}$-rank r and $X^{\{2\}}$-rank s. Note that each observation above is generated from kl simple random units independent of the others. The procedure described above can be extended to general p-layer RSS straightforwardly only with increasing complexity of notations.

The two-layer RSS is described in the way above only for the better understanding of the procedure. In implementation, the procedure can be carried out as follows. To get one measurement on Y with a specified rank pair, say, on $Y_{[r][s]}$, l sets of units, each of size k, are drawn and then ranked with respect to $X^{\{1\}}$. The l units in the ranked sets with $X^{\{1\}}$-rank r are selected and then ranked with respect to $X^{\{2\}}$. Finally, the unit with $X^{\{2\}}$-rank s is selected for the measurement of Y. It is not necessary to repeat the process the same number of times for each pair (r, s). If the process is repeated the same number of times for each pair, the two-layer RSS is balanced, otherwise, it is unbalanced. In this section, we only deal with balanced MRSS. Unbalanced

MRSS will be considered when we deal with optimal designs for regression analysis.

6.1.2 Consistency of multi-layer ranked set sampling

Let F denote the distribution function of Y. Let $F_{[r][s]}$ denote the common distribution function of the observations $Y_{[r][s]j}$ and $F_{[r]}$ denote the common distribution function of the latent values of $Y_{[r]i}$. The observation generating procedure of the two-layer RSS can be illustrated by the diagram below:

$$Y \sim F \begin{cases} Y_{[1]} \sim F_{[1]} \begin{cases} Y_{[1][1]} \sim F_{[1][1]} \\ Y_{[1][2]} \sim F_{[1][2]} \\ \cdots \\ Y_{[1][l]} \sim F_{[1][l]} \end{cases} \\ Y_{[2]} \sim F_{[2]} \begin{cases} Y_{[2][1]} \sim F_{[2][1]} \\ Y_{[2][2]} \sim F_{[2][2]} \\ \cdots \\ Y_{[2][l]} \sim F_{[2][l]} \end{cases} \\ \cdots \\ Y_{[k]} \sim F_{[k]} \begin{cases} Y_{[k][1]} \sim F_{[k][1]} \\ Y_{[k][2]} \sim F_{[k][2]} \\ \cdots \\ Y_{[k][l]} \sim F_{[k][l]} \end{cases} \end{cases} \tag{6.2}$$

In the diagram, the variables to the right of a curly bracket are generated by a marginal RSS procedure from the distribution to the left of the curly bracket. The notation $Y \sim F$ reads as Y follows distribution F. Note that $F_{[r]}$ is the distribution function of the rth $X^{\{1\}}$-induced order statistic of a random sample of size k from F, and that, for fixed r, $F_{[r][s]}$ is the distribution function of the sth $X^{\{2\}}_{[r]}$-induced order statistic of a simple random sample of size l from $F_{[r]}$, where $X^{\{2\}}_{[r]}$ is the rth $X^{\{1\}}$-induced order statistic of a random sample of size k from the distribution of $X^{\{2\}}$. It follows from the consistency of a marginal RSS that

$$F = \frac{1}{k}\sum_{r=1}^{k} F_{[r]}, \quad F_{[r]} = \frac{1}{l}\sum_{s=1}^{l} F_{[r][s]}.$$

Therefore

$$F = \frac{1}{kl}\sum_{r=1}^{k}\sum_{s=1}^{l} F_{[r][s]}. \tag{6.3}$$

That is, the two-layer RSS is consistent. In general, we can show by the similar argument that any multi-layer RSS is consistent.

The consistency of the MRSS implies, by the theory developed in Chapters 2 and 3, that an MRSS is more efficient than an SRS: the amount of useful information contained in an MRSS sample is larger, the MRSS results in a smaller variance in the estimation of means, and it results in a smaller asymptotic variance in the estimation of smooth functions of means and population quantiles, etc.

6.1.3 Comparison between multi-layer RSS and marginal RSS by simulation studies

In this sub-section, we are concerned with the comparison between the two-layer RSS and the marginal RSS, i.e., the RSS using a single concomitant variable as the ranking criterion. For the comparison to be meaningful, the two-layer RSS and the marginal RSS must be put on the same footing, that is, the sizes of the ranked sets must be determined so that the two sampling schemes result in the same number of "strata" from the stratification point of view. Therefore, the two-layer RSS with set size k in the first layer and set size l in the second layer should be compared with the marginal RSS with set size $k' = kl$. Unfortunately, a theoretical comparison between the two-layer RSS and the marginal RSS, like that between the two-layer RSS and SRS, is impossible. We make some empirical comparisons by simulation studies in this sub-section.

The following regression model is considered in our simulation studies:

$$Y = \beta_1 X_1 + \beta_2 X_2 + \epsilon, \tag{6.4}$$

where X_1 and X_2 are taken as i.i.d. standard normal variables and ϵ is a normal random error independent of X_1 and X_2. Two settings of model (6.4) is taken: (i) $\beta_1 = -\beta_2 = 2$ and (ii) $\beta_1 = \sqrt{7}, \beta_2 = -\sqrt{3}$. Under each setting, the variance of ϵ is determined by a pre-specified R^2. The values of R^2 considered are $0.6, 0.75$ and 0.9. By setting (i), we intend to compare the two-layer RSS with the marginal RSS when no concomitant variable dominates the association with the variable of interest. By setting (ii), we intend to compare the two-layer RSS with the marginal RSS when there is a dominant concomitant variable. Under setting (i), the two-layer RSS is compared with the marginal RSS using X_1 as the ranking variable. Under setting (ii), the two-layer RSS is compared with the two marginal RSS's using X_1 and X_2 as the ranking variable respectively. For the two-layer RSS, the set sizes are taken as $k = l = 3$ and, for the marginal RSS, the set size is taken as $k' = 9$. Two sample sizes, $N = 18$ and 36, corresponding to $m = 2$ and 4, are considered. The protocol of the simulation is as follows. For each setting with given R^2 and m, 2000 copies of samples with the two-layer RSS and the marginal RSS schemes are generated. The two-layer RSS sample and the marginal RSS samples in each copy are generated from common simple random samples (the sets). Based on each of these samples, the following estimators are computed: the estimator of the 0.7th quantile, the sample mean estimator and the regression estimator of the population mean, and the estimators of the regression coefficients. The latter regression estimates will be discussed in detail in Section

Table 6.1. The MSE's of the estimators based on two-layer RSS and marginal RSS samples under setting 1

m	R^2	Scheme	0.7q	$\bar{Y}_{RSS}$	$\hat{\mu}_{REG}$	$\hat{\beta}_1$	$\hat{\beta}_2$
2	0.60	Two-layer	2.353	0.533	0.356	0.363	0.351
		Marginal(X_1)	2.359	0.526	0.347	0.335	0.374
	0.75	Two-layer	1.722	0.368	0.203	0.182	0.180
		Marginal(X_1)	1.812	0.419	0.200	0.163	0.190
	0.90	Two-layer	1.475	0.284	0.098	0.060	0.058
		Marginal(X_1)	1.496	0.298	0.100	0.055	0.062
4	0.60	Two-layer	1.868	0.268	0.176	0.171	0.157
		Marginal(X_1)	1.923	0.280	0.176	0.159	0.163
	0.75	Two-layer	1.561	0.187	0.099	0.079	0.081
		Marginal(X_1)	1.564	0.206	0.102	0.080	0.083
	0.90	Two-layer	1.309	0.139	0.049	0.027	0.027
		Marginal(X_1)	1.320	0.160	0.049	0.025	0.028

6.3. For each estimator, the approximate mean square error is computed as $\text{MSE}(\hat{\theta}) = \frac{1}{2000}\sum_{j=1}^{2000}(\hat{\theta}_j - \theta)^2$ where θ is the true value of the parameter and $\hat{\theta}_j$ is the estimate based on the jth sample. The simulated MSEs for settings 1 and 2 are given, respectively, in Tables 6.1 and 6.2.

Table 6.2. The MSE's of the estimators based on two-layer RSS and marginal RSS samples under setting 2

m	R^2	Scheme	0.7q	$\bar{Y}_{RSS}$	$\hat{\mu}_{REG}$	$\hat{\beta}_1$	$\hat{\beta}_2$
2	0.60	Two-layer	2.933	0.669	0.477	0.438	0.442
		Marginal(X_1)	2.632	0.608	0.466	0.440	0.472
		Marginal(X_2)	2.984	0.784	0.450	0.473	0.428
	0.75	Two-layer	2.347	0.487	0.244	0.214	0.215
		Marginal(X_1)	2.230	0.435	0.255	0.209	0.247
		Marginal(X_2)	2.454	0.585	0.246	0.231	0.207
	0.90	Two-layer	1.866	0.360	0.131	0.075	0.074
		Marginal(X_1)	1.735	0.317	0.178	0.068	0.075
		Marginal(X_2)	1.917	0.467	0.129	0.079	0.070
4	0.60	Two-layer	2.425	0.326	0.220	0.198	0.203
		Marginal(X_1)	2.443	0.324	0.221	0.199	0.198
		Marginal(X_2)	2.511	0.420	0.233	0.228	0.200
	0.75	Two-layer	1.906	0.243	0.121	0.099	0.096
		Marginal(X_1)	1.915	0.226	0.130	0.104	0.100
		Marginal(X_2)	1.974	0.323	0.124	0.107	0.100
	0.90	Two-layer	1.599	0.181	0.061	0.032	0.033
		Marginal(X_1)	1.575	0.156	0.061	0.033	0.031
		Marginal(X_2)	1.643	0.455	0.069	0.035	0.032

The results in Table 6.1 reveal that (i) the two-layer RSS is more efficient than the marginal RSS in the estimation of the population mean and the 0.7th quantile, (ii) the regression estimate is always more efficient than the ordinary sample mean estimate, and the efficiency of the regression estimate does not seem to depend on the ranking schemes since the MSE's of the two-layer RSS and the marginal RSS in the regression estimates of the mean are comparable to each other, and (iii) the two regression coefficients β_1 and β_2 are estimated in about the same accuracy with the two-layer RSS while β_1 is estimated more accurately than β_2 with the marginal RSS using X_1 as the ranking variable.

From Table 6.2, the following can be seen: (i) in the estimation of the population mean and the 0.7th quantile, the marginal RSS using X_1 as the ranking variable is the most efficient, the marginal RSS using X_2 as the ranking variable is the least efficient, and the efficiency of the two-layer RSS is in between; (ii) in the regression estimates, the same phenomenon as in Table 6.1 reappears, that is, the regression estimate is always more efficient than the ordinary sample mean estimate and the efficiency of the regression estimate of the population mean does not seem to depend on the ranking schemes; and (iii) in the estimation of the regression coefficients, with the marginal RSS, the regression coefficient corresponding to the concomitant variable used for ranking is estimated more accurately than the other, however, with the two-layer RSS, both regression coefficients are estimated in about the same accuracy.

Now let us discuss some points indicated by the simulation studies presented above. We are dealing with three problems: (a) the estimation of the features of Y without regression, (b) the regression estimation of the mean of Y, and (c) the estimation of the regression coefficients. For problem (a), which sampling scheme is more efficient depends on which one results in a stratification such that the features of Y to be estimated are more homogeneous within each stratum. Intuitively, the strata of squares will be more homogeneous than the strata of slices unless the slices result from a concomitant variable that is strongly correlated with Y. Therefore, in general, the two-layer RSS would be more efficient than the marginal RSS unless the ranking variable in the marginal RSS is strongly correlated with Y. Since, in practice, it is usually unknown a priori whether a concomitant variable is strongly correlated with Y, the two-layer RSS is more "robust" than a marginal RSS in the sense that even if a strongly correlated concomitant variable exists the two-layer RSS does not lose too much efficiency compared with the best marginal RSS. For problem (b), we will see in Section 6.3 that the phenomenon appearing in the simulation studies is not incidental. If the ranking criterion is confined to the class of linear functions of the concomitant variables, the variance of the regression estimator of μ_Y for large m is the same regardless the particular linear function used in the ranking. Since the ranking mechanism of the two-layer RSS is essentially linear, it is not surprising that, in the simulation studies, the variances of the regression estimators do not depend on the sampling schemes. Therefore, if the regression estimator of μ_Y is the only concern, it does not matter which sampling scheme is to be used. For problem (c), however, the sampling scheme does matter. The two-layer RSS would give a balanced estimation of the regression coefficients in terms of accuracy by using the same set size in the two layers of the ranking scheme. It can also be readily adjusted to achieve different accuracy for the estimation of different coefficients by using different set sizes. The marginal RSS, however, cannot serve this purpose.

6.1.4 Issues on the choice of concomitant variables

When there are many concomitant variables, we need to make a selection as to which ones are to be used in the MRSS. We make a brief discussion on this issue in this sub-section.

The principle for the selection of concomitant variables is similar to that for the choice of independent variables in multiple regression: only variables that are significantly correlated with the variable of interest should be selected; variables, which are selected, should not exhibit strong collinearity. But, by the nature of the usage of the concomitant variables in RSS, the implementation of the selection differs from that in multiple regression — a choice of the concomitant variables must be made before measurements on the response variable are taken. In the following, we distinguish the selection based on the significance of the concomitant variables and that based on the collinearity among the concomitant variables.

Preliminary selection based on collinearity. The collinearity among the concomitant variables can be analyzed using the information on the concomitant variables alone. Since, by assumption, the values of the concomitant variables can be easily obtained with negligible cost, a preliminary sampling on the values of the concomitant variables can be conducted and the collinearity among them can then be analyzed. When several variables exhibit a strong collinearity, only one of them should be selected for the MRSS procedure. The rationale behind this rule is heuristically argued as follows. Consider an extreme case. Suppose two concomitant variables are almost perfectly correlated. The question whether to select only one of them or not is then equivalent to whether to rank the sampling units based on a single variable in the following one way or another: (i) rank a set of k^2 sampling units all together according to the values of the variable, (ii) rank the k^2 units in the manner of a two-layer RSS but use the same variable as the ranking variable at both layers. Suppose the ranking variable is the response variable itself and we are concerned with the estimation of the mean of this variable. The first way of ranking then results in a relative efficiency (with respect to SRS), roughly, $(k^2+1)/2$. The second way of ranking, however, results in a relative efficiency, roughly, $(k+1)^2/4$. The first way of ranking is always more efficient than the second way of ranking.

Selection based on the significance of the concomitant variables. Before any measurements on the response variable are taken, the assessment of the significance of a concomitant variable in its relation to the response variable might be made by using some prior information, if available, or by guess work. An objective assessment of the significance can be made after some measurements on the response variable are obtained. The following dynamic version of the multi-layer RSS can be applied in practice. In the earlier stage of the sampling, more concomitant variables, which are potentially significant, are used in the ranking mechanism. Once enough measurements on the response variable are made, the significance of each concomitant variable is assessed. Then, discard the non-significant concomitant variables and use the remaining ones in the ranking mechanism. As long as the ranking mechanism is the same at each cycle of the RSS, the resultant sampling is at least more efficient than SRS.

6.2 Adaptive ranked set sampling

If only the estimation of the features of Y is of concern, the multi-layer RSS is not necessarily more efficient than a particular marginal RSS in certain situations, as discussed in the last section. Is there a ranking mechanism that is better than any marginal RSS in any situation? Or, more generally, is there a most efficient ranking mechanism? In this section, we answer these questions and develop an adaptive RSS scheme based on the ranking mechanism that is the best among those determined by functions of concomitant variables.

6.2.1 The best ranking mechanism

Suppose an RSS is carried out with set size k in m cycles. Recall that, for any consistent ranking mechanism, we can express the distribution function of the rth ranked statistic of Y as $F_{[r]} = \sum_{s=1}^{k} p_{rs} F_{(s)}$, where the p_{rs}'s satisfy $\sum_{s=1}^{k} p_{rs} = 1$ and $\sum_{r=1}^{k} p_{rs} = 1$, and $F_{(s)}$ is the distribution of the sth (exact) order statistic of Y, see Lemma 3.6.

Now consider the estimation of the mean μ_Y of Y. Denote the RSS sample mean with an arbitrary consistent ranking mechanism by $\bar{Y}_{[\text{RSS}]} = 1/(mk) \sum_{r=1}^{k} \sum_{j=1}^{m} Y_{[r]j}$, which is unbiased for the estimation of μ_Y. We have

$$\text{Var}(\bar{Y}_{[\text{RSS}]}) = \frac{1}{mk}[m_2 - \frac{1}{k}\sum_{r=1}^{k}(EY_{[r]})^2],$$

where m_2 is the second moment of Y. Now, by the expression of $F_{[r]}$,

$$\begin{aligned}\sum_{r=1}^{k}(EY_{[r]})^2 &= \sum_{r=1}^{k}(\sum_{s=1}^{k} p_{rs} EY_{(s)})^2 \leq \sum_{r=1}^{k}\sum_{s=1}^{k} p_{rs}(EY_{(s)})^2 \\ &= \sum_{s=1}^{k}(\sum_{r=1}^{k} p_{rs})(EY_{(s)})^2 = \sum_{s=1}^{k}(EY_{(s)})^2.\end{aligned}$$

Denote by $\bar{Y}_{(\text{RSS})}$ the sample mean of a perfect RSS with the same set size and number of cycles. The results above imply that $\text{Var}(\bar{Y}_{[\text{RSS}]}) \geq \text{Var}(\bar{Y}_{(\text{RSS})})$, that is, the perfect ranking mechanism is the best in the estimation of μ_Y.

Similarly, we can show that the perfect ranking mechanism is the best for the estimation of the mean of any function of Y, of any smooth-function-of-means, of population quantiles, etc. In summary, the perfect ranking mechanism is the most efficient among all ranking mechanisms.

As an implicit assumption in our context, however, the ranking of the sampling units cannot be done with respect to Y without the actual measurement of Y, let alone perfect ranking. Since the concomitant variables are the only resource for the consideration of ranking mechanisms, we consider the ranking mechanisms that rank the sampling units according to the values of a function of the concomitant variables. Obviously, the best function is the one that has the highest correlation with Y — the more a function is correlated with Y the more the ranking mechanism is like perfect ranking. Let $g(\boldsymbol{X})$ be any function of $\boldsymbol{X}$. Denote by ρ_{Yg} the correlation between Y and $g(\boldsymbol{X})$. We have

$$\begin{aligned}
\rho_{Yg}^2 &= \left[\frac{E(Y-\mu_Y)(g(\boldsymbol{X})-\mu_g)}{\sigma_Y\sigma_g}\right]^2 \\
&= \left[\frac{E[E[(Y-\mu_Y)(g(\boldsymbol{X})-\mu_g)|\boldsymbol{X}]]}{\sigma_Y\sigma_g}\right]^2 \\
&= [E[\tilde{g}(\boldsymbol{X})(E[Y|\boldsymbol{X}]-\mu_Y)/\sigma_Y]]^2 \\
&\leq E[\tilde{g}(\boldsymbol{X})^2]E[(E[Y|\boldsymbol{X}]-\mu_Y)/\sigma_Y]^2 \\
&= E[(E[Y|\boldsymbol{X}]-\mu_Y)/\sigma_Y]^2,
\end{aligned}$$

where μ_Y, σ_Y and μ_g, σ_g are the means and standard deviations of Y and $g(\boldsymbol{X})$ respectively, and $\tilde{g}(\boldsymbol{X}) = [g(\boldsymbol{X}) - \mu_g]/\sigma_g$. The equal sign in the inequality above holds if and only if $\tilde{g}(\boldsymbol{X})$ is a linear function of $E[Y|\boldsymbol{X}]$. Therefore, we conclude that the function which has the highest correlation with Y is the conditional expectation $E[Y|\boldsymbol{X}]$.

6.2.2 Adaptive ranked set sampling procedure

Denote the conditional expectation $E[Y|\boldsymbol{X}]$ by $g(\boldsymbol{X})$. The conditional expectation, however, cannot be readily used for ranking since it is unknown. To cope with this difficulty, we propose an adaptive RSS procedure. The procedure is a cyclical process described as follows. At an initial stage, RSS is conducted using any reasonable ranking mechanism, a marginal RSS ranking mechanism or a multi-layer RSS ranking mechanism. Then the process goes in cycles of an estimation step and a sampling step. In the estimation step, the available data is used to estimate (or update the estimate of) $g(\boldsymbol{X})$. In the second step, the updated estimate of $g(\boldsymbol{X})$ is used as the ranking function for the RSS and more data are collected. The procedure is described in more detail in the following algorithm.

Table 6.3. Simulated mean square errors and relative efficiencies of Example 1 where the concomitant variables follow a joint normal distribution.

$\sigma(R^2)$	m	MSE_g	MSE_X	$\text{MSE}_{\hat{g}}$	$\text{RE}_{\hat{g}\cdot g}$	$\text{RE}_{\hat{g}\cdot X}$
	4	0.2182	0.3951	0.2889	0.7553	1.3676
1	8	0.1102	0.1947	0.1298	0.8490	1.5000
(0.90)	16	0.0545	0.0891	0.0566	0.9629	1.5742
	4	0.5758	0.7148	0.6520	0.8831	1.0963
2.82	8	0.2886	0.3741	0.3197	0.9027	1.1702
(0.54)	16	0.1407	0.1875	0.1492	0.9430	1.2567
	4	3.4911	3.5364	3.4796	1.0033	1.0163
8	8	1.5995	1.7336	1.6447	0.9725	1.0541
(0.13)	16	0.8240	0.8961	0.8356	0.9861	1.0724

Initial stage: *Choose a reasonable ranking mechanism. For* $r = 1, \ldots, k$, *draw a random sample of* k *units from the population, rank the units by the chosen mechanism, and then measure the* Y *value for the unit ranked* r. *Repeat this cycle a number of times until the function* $g(\boldsymbol{X})$ *can be estimated reasonably well. Denote the data so obtained by* $DATA_0$.

Recursive cycles: *For* $j = 1, 2, \ldots$

(i) Compute the estimator $\hat{g}^{(j)}$ *of* g *using* $DATA_{j-1}$.

(ii) For $r = 1, \ldots, k$, *draw a random sample of* k *units from the population, rank the units according to the orders of the* $\hat{g}^{(j-1)}(\boldsymbol{X})$ *values, and then measure the* Y *value for the unit ranked* r.

(iii) Augment $DATA_{j-1}$ *to* $DATA_j$.

For the purpose of ranking, the function g does not need to be estimated very accurately. As long as the estimate becomes direction-stable in the sense that the sign of the partial derivatives of the estimated g stay unchanged over most of the range of $\boldsymbol{X}$. As a consequence, the initial stage does not need to take many cycles. When the sample size gets larger, the efficiency of the procedure approaches the best possible — the estimate of g then converges to g.

6.2.3 A simulation study

We present the results of a simulation study in this sub-section to illustrate the efficiency of the adaptive RSS procedure. In the simulation study, the following model is considered in three examples: $Y = g(\boldsymbol{X}) + \epsilon$, where $\boldsymbol{X}$ is a random vector independent of ϵ. In examples 1 and 2, $g(\boldsymbol{X}) = -2X_1 + 2X_2 + 2X_3$, while in Example 1, X_1, X_2, X_3 follow a normal distribution $N_3(0, 0.75I + 0.35\mathbf{1}\mathbf{1}')$, and in Example 2, X_1, X_2, X_3 are i.i.d. with an uniform distribution on interval $[-2,\ 2]$. In example 3, $g(X) = 2\sin(X) + 3\cos(X)$, and X is

Table 6.4. Simulated mean square errors and relative efficiencies of Example 2 where the concomitant variables are i.i.d. with a uniform distribution.

$\sigma(R^2)$	m	MSE_g	MSE_X	$MSE_{\hat{g}}$	$RE_{\hat{g}\cdot g}$	$RE_{\hat{g}\cdot X}$
	4	0.3324	0.6627	0.4511	0.7369	1.4691
1	8	0.1533	0.3258	0.1797	0.8531	1.8130
(0.94)	16	0.0801	0.1731	0.0888	0.9020	1.9493
	4	0.6400	1.0020	0.8252	0.7756	1.2143
2.82	8	0.3407	0.5168	0.3774	0.9028	1.3694
(0.67)	16	0.1680	0.2505	0.1844	0.9111	1.3585
	4	3.4582	3.9526	3.6442	0.9490	1.0846
8	8	1.7895	1.8579	1.8048	0.9915	1.0294
(0.2)	16	0.8864	1.0120	0.9522	0.9309	1.0628

Table 6.5. Simulated mean square errors and relative efficiencies of Example 3 where the regression function is nonlinear.

σ	m	MSE_g	MSE_X	$MSE_{\hat{g}}$	$RE_{\hat{g}\cdot g}$	$RE_{\hat{g}\cdot X}$
	4	0.1523	0.3162	0.1968	0.7739	1.6067
1	8	0.0947	0.2223	0.1167	0.8115	1.9049
	16	0.0754	0.1597	0.0879	0.8578	1.8168
	4	0.3389	0.5038	0.3958	0.8562	1.2729
2.82	8	0.2133	0.3247	0.2327	0.9166	1.3954
	16	0.1671	0.2499	0.1845	0.9057	1.3545
	4	1.7339	1.9451	1.8874	0.9187	1.0306
8	8	1.2108	1.2830	1.1719	1.0332	1.0948
	16	0.7771	0.9311	0.8904	0.8728	1.0457

uniformly distributed on interval $[0, 2\pi]$. In all the examples, ϵ has a normal distribution $N(0, \sigma^2)$ with varying σ. The simulation is carried out as follows. The initial stage of the adaptive RSS takes a single cycle and a single X is used for ranking. For later cycles, the ranking mechanism of the adaptive RSS is based on the estimated g. In order to assess the efficiency of the adaptive RSS, we consider, at the same time, a marginal RSS and the RSS using the true function g as the ranking function. In each cycle of the simulated RSS procedure, for $r = 1, \cdots, k$, a simple random sample of $(Y, \boldsymbol{X})$ of size k is generated, then the sample is, respectively, ranked according to $g(\boldsymbol{X})$, a pre-specified single X and $\hat{g}(\boldsymbol{X})$, the $(Y, \boldsymbol{X})$ value of the unit with rank r corresponding to each of the three ranking mechanisms is recorded. This cycle is repeated m times, generating three ranked set samples, each of size $n = km$. In Examples 1 and 2, the single X is taken as X_3 and the linear function is estimated using Splus function *lm*. In Example 3, the function g is estimated using Splus function *smooth.spline*. In all the examples, $k = 5$, $m = 4, 8, 16$, and $\sigma = 1, 2.82, 8$. The σ values determine different values of R^2 defined as the ratio $\text{Var}(g(\boldsymbol{X}))/\text{Var}(Y)$. For each example, 2000 independent copies of the ranked set samples as described above are generated. For each sample, the sample mean is computed as the estimate of the population mean. The mean square error over 2000 samples are computed and presented in Tables 6.3—6.5. In the tables, MSE_g, MSE_X and $\text{MSE}_{\hat{g}}$ denote, respectively, the mean square errors of the RSS estimates with ranking criteria g, X and $\hat{g}$, $\text{RE}_{\hat{g}\cdot g}$ and $\text{RE}_{\hat{g}\cdot X}$ are defined as $\text{MSE}_g/\text{MSE}_{\hat{g}}$ and $\text{MSE}_X/\text{MSE}_{\hat{g}}$, respectively.

From the tables, we can observe the following features. (i) The adaptive procedure is always better than the marginal RSS, and the efficiency gain is very significant in certain cases. (ii) The efficiency of the adaptive RSS approaches rapidly that of the RSS with the ideal ranking function g as the sample size gets larger. In the case $\sigma = 8$, the two values of $\text{RE}_{\hat{g}\cdot g}$, which are bigger than 1, are explicable by the fact that the R^2 is too small and the model is too erratic.

6.3 Regression analysis based on RSS with concomitant variables

Suppose that the variable Y and the concomitant variables $\boldsymbol{X}$ follow a linear regression model, i.e.,

$$Y = \alpha + \boldsymbol{\beta}'\boldsymbol{X} + \epsilon, \tag{6.5}$$

where $\boldsymbol{\beta}$ is a vector of unknown constant coefficients, and ϵ is a random variable with mean zero and variance σ_ϵ^2 and is independent of $\boldsymbol{X}$.

Suppose that an RSS with certain ranking mechanism is implemented in m cycles. In a typical cycle i, for $r = 1, \ldots, k$, a simple random sample of k units with latent values $(Y_{1ri}, \boldsymbol{X}_{1ri}), \cdots, (Y_{kri}, \boldsymbol{X}_{kri})$ is drawn from the population. The values of the $\boldsymbol{X}$'s are all measured. The k sampled units are ranked according to the ranking mechanism. Then the Y value of the unit with rank r is measured. The $\boldsymbol{X}$ and Y values of the unit with rank r are denoted, respectively, by $\boldsymbol{X}_{[r]i}$ and $Y_{[r]i}$. In the completion of the sampling, we have a data set as follows:

$$\begin{array}{c}(Y_{[1]i}, \boldsymbol{X}_{[1]i}, \boldsymbol{X}_{11i}, \ldots, \boldsymbol{X}_{1ki}),\\ \cdots \\ (Y_{[k]i}, \boldsymbol{X}_{[k]i}, \boldsymbol{X}_{k1i}, \ldots, \boldsymbol{X}_{kki}),\\ i = 1, \ldots, m.\end{array} \tag{6.6}$$

Note that the setting above includes both the multi-layer RSS and the adaptive RSS discussed in the previous sections. In fact, we can treat the ranks as indices of strata or categories. Thus, in the two-layer RSS, we can represent the double ranks by a single index in an appropriate manner. Based on the data set (6.6), we consider in this section two problems: (i) the estimation of the regression coefficients, (ii) the estimation of μ_Y, the mean of Y.

6.3.1 Estimation of regression coefficients with RSS

Denote by $G(y|\boldsymbol{x})$ the conditional distribution of Y given $\boldsymbol{X}$. Let $\{(Y_1, \boldsymbol{X}_1), \ldots, (Y_n, \boldsymbol{X}_n)\}$ be a simple random sample from the joint distribution of Y and $\boldsymbol{X}$. Suppose $(R_1, \ldots, R_n)$ is a random permutation of $(1, \ldots, n)$ determined only by $(\boldsymbol{X}_1, \ldots, \boldsymbol{X}_n)$. We can extend a result by Bhattacharya ([18], Lemma3.1) to the following:

Lemma 6.1. *The random-permutation-induced statistics $Y_{R_1}, \ldots, Y_{R_n}$ are conditionally independent given $(\boldsymbol{X}_1, \ldots, \boldsymbol{X}_n)$ with conditional distribution functions $G(\cdot|X_{R_1}), \ldots, G(\cdot|X_{R_n})$ respectively.*

As remarked by Bhattacharya ([18],page 385), although his lemma deals only with the induced order statistics, the key fact on which the proof of that lemma depends is that the induced orders are a random permutation of $1, \ldots, n$ determined only by the X's. Therefore, the extension above is trivial. Note that the ranking mechanism in RSS indeed produces a random permutation of the sampling units. Hence the lemma above can be applied to the RSS sample. In particular, we have

$$Y_{[r]i} = \alpha + \boldsymbol{\beta}^T \boldsymbol{X}_{[r]i} + \epsilon_{ri}, \qquad (6.7)$$
$$r = 1, \ldots, k; i = 1, \ldots, m,$$

where ϵ_{ri} are independent identically distributed as the ϵ in (6.5) and are independent of $\boldsymbol{X}_{[r]i}$. Thus, we can estimate α and $\boldsymbol{\beta}$ by least squares method based on (6.7).

Let

$$\bar{\boldsymbol{X}}_{\text{RSS}} = \frac{1}{mk}\sum_{r=1}^{k}\sum_{i=1}^{m}\boldsymbol{X}_{[r]i}, \quad \bar{Y}_{\text{RSS}} = \frac{1}{mk}\sum_{r=1}^{k}\sum_{i=1}^{m}Y_{[r]i},$$
$$X_{\text{RSS}} = (\boldsymbol{X}_{[1]1}, \ldots, \boldsymbol{X}_{[1]m}, \ldots, \boldsymbol{X}_{[k]1}, \ldots, \boldsymbol{X}_{[k]m})'$$
$$\boldsymbol{Y}_{\text{RSS}} = (y_{[1]1}, \ldots, y_{[1]m}, \ldots, y_{[k]1}, \ldots, y_{[k]m})'.$$

The least squares estimates of α and $\boldsymbol{\beta}$ based on (6.7) are then given, respectively, by

$$\hat{\alpha}_{\text{RSS}} = \bar{Y}_{\text{RSS}} - \hat{\boldsymbol{\beta}}'_{\text{RSS}}\bar{\boldsymbol{X}}_{\text{RSS}}, \qquad (6.8)$$

$$\hat{\boldsymbol{\beta}}_{\text{RSS}} = [X'_{\text{RSS}}(I - \frac{\mathbf{11}'}{mk})X_{\text{RSS}}]^{-1}X'_{\text{RSS}}(I - \frac{\mathbf{11}'}{mk})\boldsymbol{Y}_{\text{RSS}}. \qquad (6.9)$$

It is obvious that $\hat{\alpha}_{\text{RSS}}$ and $\hat{\boldsymbol{\beta}}_{\text{RSS}}$ are unbiased. Since both $\hat{\alpha}_{\text{RSS}}$ and $\hat{\boldsymbol{\beta}}_{\text{RSS}}$ are of the form of smooth-function-of-means, as estimates of α and $\boldsymbol{\beta}$, they are, at least asymptotically, as good as their counterparts based on an SRS, that is, their asymptotic variances at the order $O(\frac{1}{mk})$ are smaller than or equal to those based on an SRS. More specific conclusions can be made from the explicit expressions of the variances of $\hat{\alpha}_{\text{RSS}}$ and $\hat{\boldsymbol{\beta}}_{\text{RSS}}$. These variances can be derived as

$$\text{Var}(\hat{\alpha}_{\text{RSS}}) = \sigma_\epsilon^2 E\left[\frac{1}{mk} + \bar{\boldsymbol{X}}_{\text{RSS}}^T[X_{\text{RSS}}'(I - \frac{\mathbf{11}'}{mk})X_{\text{RSS}}]^{-1}\bar{\boldsymbol{X}}_{\text{RSS}}\right],$$

$$\text{Var}(\hat{\boldsymbol{\beta}}_{\text{RSS}}) = \sigma_\epsilon^2 E\left[[X_{\text{RSS}}'(I - \frac{\mathbf{11}'}{mk})X_{\text{RSS}}]^{-1}\right],$$

where the expectations are taken with respect to the distribution of the X's. Let Σ denote the variance-covariance matrix of $\boldsymbol{X}$. We have

$$\frac{1}{mk}X_{\text{RSS}}'(I - \frac{\mathbf{11}'}{mk})X_{\text{RSS}} \to \Sigma,$$

almost surely. In order to get the asymptotic expressions of the variances above at the order $O(\frac{1}{mk})$, we can replace $[X_{\text{RSS}}'(I - \frac{\mathbf{11}'}{mk})X_{\text{RSS}}]^{-1}$ by $\frac{1}{mk}\Sigma^{-1}$. Then, it is easy to see that the asymptotic variances at the order $O(\frac{1}{mk})$ are the same as those of their counterparts based on an SRS, which implies that, in the estimation of the regression coefficients, balanced RSS and SRS are asymptotically equivalent. Balanced RSS cannot do much for the improvement of the estimation of the regression coefficients. If the estimation of these coefficients is the main concern, we need to consider unbalanced RSS which will be discussed in detail in Section 6.4.

6.3.2 Regression estimate of the mean of Y with RSS

Let

$$\bar{\boldsymbol{X}}_T = \frac{1}{mk^2}\sum_{r=1}^{k}\sum_{j=1}^{k}\sum_{i=1}^{m}\boldsymbol{X}_{rji}.$$

We can define another estimate of μ_Y rather than $\bar{Y}_{\text{RSS}}$. The estimate is called the RSS regression estimate and is defined as

$$\hat{\mu}_{\text{RSS}\cdot\text{REG}} = \bar{Y}_{\text{RSS}} + \hat{\boldsymbol{\beta}}_{\text{RSS}}'(\bar{X}_T - \bar{X}_{\text{RSS}}). \tag{6.10}$$

The RSS regression estimate of μ_Y is unbiased and its variance can be obtained as

$$\text{Var}(\hat{\mu}_{\text{RSS}\cdot\text{REG}}) = \frac{\sigma_\epsilon^2}{mk}\{1 + \Delta_{\text{RSS}}\} + \frac{1}{mk^2}\boldsymbol{\beta}'\Sigma\boldsymbol{\beta}, \tag{6.11}$$

where

$$\Delta_{\text{RSS}} = E[mk(\bar{X}_T - \bar{X}_{\text{RSS}})'[X_{\text{RSS}}'(I - \frac{\mathbf{11}'}{mk})X_{\text{RSS}}]^{-1}(\bar{X}_T - \bar{X}_{\text{RSS}})].$$

If the ranked set sample is replaced by a simple random sample, we get an SRS regression estimate of μ_Y. The variance of the SRS regression estimate is also of the form (6.11) but with Δ_{RSS} replaced by the corresponding quantity

Δ_{SRS} defined on the simple random sample. It follows from the asymptotic approximations of Δ_{RSS} and Δ_{SRS}, which will be given later, that, as long as the ranking mechanism in the RSS is consistent, we always have $\Delta_{\text{RSS}} < \Delta_{\text{SRS}}$ asymptotically. In other words, the RSS regression estimate is asymptotically more efficient than the SRS regression estimate.

6.3.3 Comparison of the RSS regression estimate and the ordinary RSS estimate

We now have two unbiased estimates of μ_Y: the ordinary RSS estimate $\bar{Y}_{\text{RSS}}$ and the RSS regression estimate $\hat{\mu}_{\text{RSS}\cdot\text{REG}}$. Which of these two estimates is better? We need to compare their variances.

Let $\Sigma_{[r]}$ denote the variance-covariance matrix of $\boldsymbol{X}_{[r]}$, the $\boldsymbol{X}$ value of the unit with rank r. Let $\Sigma_{\text{RSS}} = \frac{1}{k}\sum_{r=1}^{k}\Sigma_{[r]}$. It can be easily derived that

$$\text{Var}(\bar{Y}_{\text{RSS}}) = \frac{1}{mk}\boldsymbol{\beta}'\Sigma_{\text{RSS}}\boldsymbol{\beta} + \frac{1}{mk}\sigma_\epsilon^2. \tag{6.12}$$

An approximation to Δ_{RSS} can be obtained as follows.

$$\begin{aligned}\Delta_{\text{RSS}} &\approx E[(\bar{\boldsymbol{X}}_T - \bar{\boldsymbol{X}}_{\text{RSS}})'\Sigma^{-1}(\bar{\boldsymbol{X}}_T - \bar{\boldsymbol{X}}_{\text{RSS}})] \\ &= E[tr[\Sigma^{-1}(\bar{\boldsymbol{X}}_T - \bar{\boldsymbol{X}}_{\text{RSS}})(\bar{\boldsymbol{X}}_T - \bar{\boldsymbol{X}}_{\text{RSS}})']] \\ &= tr[\Sigma^{-1}\text{Var}[(\bar{\boldsymbol{X}}_T - \bar{\boldsymbol{X}}_{\text{RSS}})]]. \end{aligned} \tag{6.13}$$

In the appendix of this chapter, it is shown that

$$\begin{aligned}\text{Var}[(\bar{\boldsymbol{X}}_T - \bar{\boldsymbol{X}}_{\text{RSS}})] &= \text{Var}[\bar{\boldsymbol{X}}_{\text{RSS}}] - \frac{1}{mk^2}\Sigma \\ &= \frac{1}{mk}\Sigma_{\text{RSS}} - \frac{1}{mk^2}\Sigma. \end{aligned} \tag{6.14}$$

Thus, we have

$$\Delta_{\text{RSS}} \approx \frac{1}{mk}[tr(\Sigma^{-1/2}\Sigma_{\text{RSS}}\Sigma^{-1/2}) - \frac{p}{k}],$$

where p is the dimension of $\boldsymbol{X}$. A similar approximation can be obtained for Δ_{SRS} with Σ_{RSS} replaced by Σ, that is,

$$\Delta_{\text{SRS}} \approx \frac{1}{mk}[tr(\Sigma^{-1/2}\Sigma\Sigma^{-1/2}) - \frac{p}{k}].$$

Since $\Sigma_{\text{RSS}} < \Sigma$, it follows that $\Delta_{\text{RSS}} < \Delta_{\text{SRS}}$ asymptotically. Now we can write the variance of $\hat{\mu}_{\text{RSS}\cdot\text{REG}}$ as

$$\text{Var}(\hat{\mu}_{\text{RSS}\cdot\text{REG}}) \approx \frac{\sigma_\epsilon^2}{mk}[1 + \frac{1}{mk}(tr\Sigma^{-1/2}\Sigma_{\text{RSS}}\Sigma^{-1/2} - \frac{p}{k})] + \frac{1}{mk^2}\boldsymbol{\beta}'\Sigma\boldsymbol{\beta}.$$

Let σ_Y^2 denote the variance of Y. We have that $\sigma_Y^2 = \boldsymbol{\beta}'\Sigma\boldsymbol{\beta} + \sigma_\epsilon^2$. Let $R^2 = \boldsymbol{\beta}'\Sigma\boldsymbol{\beta}/\sigma_Y^2$, the ratio of the variance of $\boldsymbol{\beta}'\boldsymbol{X}$ and the variance of Y. Let

$$C_{\text{RSS}}(m) = \frac{1}{m}(tr\Sigma^{-1/2}\Sigma_{\text{RSS}}\Sigma^{-1/2} - \frac{p}{k}) \text{ and } R^2_{\text{RSS}} = \frac{\boldsymbol{\beta}'\Sigma_{\text{RSS}}\boldsymbol{\beta}}{\sigma_Y^2}.$$

Using these notations, the variances of $\bar{Y}_{\text{RSS}}$ and $\hat{\mu}_{\text{RSS}\cdot\text{REG}}$ are expressed as follows:

$$\text{Var}(\bar{Y}_{\text{RSS}}) = \frac{\sigma_Y^2}{mk}(R^2_{\text{RSS}} + 1 - R^2),$$
$$\text{Var}(\hat{\mu}_{\text{RSS}\cdot\text{REG}}) \approx \frac{\sigma_Y^2(1-R^2)}{mk}[1 + \frac{1}{k}C_{\text{RSS}}(m)] + \frac{\sigma_Y^2 R^2}{mk^2}.$$

Therefore, we have the following equivalence relations:

$$\begin{aligned} &\text{Var}(\hat{\mu}_{\text{RSS}\cdot\text{REG}}) \le \text{Var}(\bar{Y}_{\text{RSS}}) \\ \Leftrightarrow\ &R^2(1 - C_{\text{RSS}}(m)) \le kR^2_{\text{RSS}} - C_{\text{RSS}}(m) \\ \Leftrightarrow\ &R^2 \le \frac{kR^2_{\text{RSS}} - C_{\text{RSS}}(m)}{1 - C_{\text{RSS}}(m)}, \text{ if } m \ge \frac{p(k-1)}{k}. \end{aligned} \tag{6.15}$$

Note that $C_{\text{RSS}}(m) < 1$ if $m \ge \frac{p(k-1)}{k}$, since

$$tr\Sigma^{-1/2}\Sigma_{\text{RSS}}\Sigma^{-1/2} \le tr\Sigma^{-1/2}\Sigma\Sigma^{-1/2} = p.$$

As $m \to \infty$, $C_{\text{RSS}}(m) \to 0$. Thus if m is large or moderate, we can ignore the term $C_{\text{RSS}}(m)$ in (6.15). The equivalence relation then becomes $R^2 \le kR^2_{\text{RSS}}$ that is equivalent to $\Sigma \le k\Sigma_{\text{RSS}}$. The latter relationship is always true, since $\Sigma > k\Sigma_{\text{RSS}}$ implies that an RSS sample of size k contains more information than the simple random sample of size k^2 from which the RSS sample is generated, which is impossible. If we drop the divisor $1 - C_{\text{RSS}}(m)$ in (6.15), we obtain a sufficient condition for $\text{Var}(\hat{\mu}_{\text{RSS}\cdot\text{REG}}) \le \text{Var}(\bar{Y}_{\text{RSS}})$: $R^2 \le kR^2_{\text{RSS}} - C_{\text{RSS}}(m)$. Unless R^2 is very small and hence R^2_{RSS} is small such that $C_{\text{RSS}}(m)$ is not negligible relative to kR^2_{RSS}, the condition will hold. Now, we can draw a general conclusion: if either m or R^2 is large, the RSS regression estimate has a smaller variance than the ordinary RSS estimate, i.e., the RSS regression estimate is better than the ordinary RSS estimate.

To shed light on how large R^2 should be for the RSS regression estimate to be better, let us consider the special case that the concomitant variables follow a joint normal distribution and the ranking criterion is $\boldsymbol{\beta}'\boldsymbol{X}$, which is the ideal ranking function. While $\text{Var}(\bar{Y}_{\text{RSS}})$ with this ranking function is the smallest among all linear ranking functions, $\text{Var}(\hat{\mu}_{\text{RSS}\cdot\text{REG}})$ is essentially the same with any linear ranking function, as will be shown in the next sub-section. It will also be shown in the next sub-section that, in this special case,

$$\Sigma_{\text{RSS}} = \Sigma - \frac{\Sigma\boldsymbol{\beta}\boldsymbol{\beta}'\Sigma}{\boldsymbol{\beta}'\Sigma\boldsymbol{\beta}}D(k).$$

Here

$$D(k) = 1 - \frac{1}{k}\sum_{r=1}^{k} \sigma_{(r)}^2,$$

where $\sigma_{(r)}^2$ is the variance of the rth order statistic of a sample of size k from the standard normal distribution. Therefore, the equivalence relationship becomes

$$\begin{aligned} &\mathrm{Var}(\hat{\mu}_{\mathrm{RSS \cdot REG}}) \le \mathrm{Var}(\bar{Y}_{\mathrm{RSS}}) \\ \Leftrightarrow\ & R^2 \ge \frac{p - k/(k-1)D(k)}{p + mk - k(mk+1)/(k-1)D(k)}. \end{aligned} \tag{6.16}$$

Let $k = 5, m = 10$. By using Table 3 in Krishnaiah and Sen [81], we obtain $D(5) = 0.6390$. For $p = 2, 3$ and 4, the values of the right hand side of inequality (6.16) are, respectively, 0.1067, 0.1795 and 0.2414.

6.3.4 How does a specific ranking mechanism affect the efficiency of the regression estimate?

The quantity Δ_{RSS} is the only component in the variance of $\hat{\mu}_{\mathrm{RSS \cdot REG}}$ which involves the ranked statistics. In this sub-section, we further study the properties of Δ_{RSS} to see how it is affected by a specific ranking mechanism.

Without loss of generality, assume that the components of $\boldsymbol{X}$ are standardized, i.e., they have mean zero and standard deviation one. We only consider ranking mechanisms determined by linear functions of $\boldsymbol{X}$ of the form $g = \boldsymbol{l}'\boldsymbol{X}$ such that $\boldsymbol{l}'\Sigma\boldsymbol{l} = 1$. To make the problem tractable, assume that $\boldsymbol{X}$ follows a multivariate normal distribution so that we can express $\boldsymbol{X}$ as

$$\boldsymbol{X} = \boldsymbol{\rho} g + \boldsymbol{\epsilon},$$

where $\boldsymbol{\rho}$ is the vector of the covariances between the component of $\boldsymbol{X}$ and g, and $\boldsymbol{\epsilon}$ follows a multivariate normal distribution with mean zero and covariance matrix $\Sigma - \boldsymbol{\rho}\boldsymbol{\rho}'$. Thus, we have

$$\boldsymbol{X}_{[r]} = \boldsymbol{\rho} g_{(r)} + \boldsymbol{\epsilon}_r, \quad r = 1, \ldots, k,$$

where $\boldsymbol{\epsilon}_r$'s are independent identically distributed as $\boldsymbol{\epsilon}$. Let $\Sigma_{[r]}$ denote the covariance matrix of $\boldsymbol{X}_{[r]}$. Then,

$$\Sigma_{[r]} = \boldsymbol{\rho}\boldsymbol{\rho}'\sigma_{(r)}^2 + \Sigma - \boldsymbol{\rho}\boldsymbol{\rho}',$$

where $\sigma_{(r)}$ is the standard deviation of the rth order statistic of a standard normal sample of size k. Notice that

$$\rho_r = \mathrm{Cov}(X_r, \boldsymbol{l}'\boldsymbol{X}) = \boldsymbol{l}'\boldsymbol{\gamma}_r,$$

where $\boldsymbol{\gamma}_r$ is the rth column of Σ. Thus,

$$\rho' = l'\Sigma.$$

Hence,

$$\Sigma_{\text{RSS}} = \Sigma - \Sigma l' l \Sigma D(k).$$

Substituting the expression above into Δ_{RSS}, we get

$$\begin{aligned}\Delta_{\text{RSS}} &= \frac{1}{mk}[tr(I_p - ll'\Sigma D(k)) - \frac{p}{k}] \\ &= \frac{1}{mk}[\frac{p(k-1)}{k} - D(k)].\end{aligned}$$

It turns out that Δ_{RSS} does not depend on the linear ranking mechanism. But this is not really surprising. It takes only a little thought to recognize the fact that, for whatever $\boldsymbol{l}$, $\boldsymbol{l}'\boldsymbol{X}$ is perfectly correlated with $\boldsymbol{X}$ (in the sense that it is completely determined by $\boldsymbol{X}$), hence, while it increases the accuracy of the ranking of certain components of $\boldsymbol{X}$, it reduces the accuracy of the others. The improvement on the estimation of μ_Y by using the RSS regression estimate is achieved through the improvement on the estimation of the regression coefficients $\boldsymbol{\beta}$. It is well comprehensible that the total improvement on the estimation of $\boldsymbol{\beta}$ should not depend on the ranking criteria such as $\boldsymbol{l}'\boldsymbol{X}$. Although, in the derivation of the result above, the normality assumption is made, we can reasonably guess that it should also be the case when the assumption is dropped, which is demonstrated by simulation studies.

6.4 Optimal RSS schemes for regression analysis

We consider again the regression model (6.5) introduced in Section 6.3. However, our focus in this section is on the estimation of the regression coefficients. As pointed out in Section 6.3, a balanced RSS and an SRS are equivalent in terms of the asymptotic variance in the estimation of the regression coefficients. In this section, we deal with unbalanced RSS schemes under certain optimality considerations so that the SRS estimate of the regression coefficients can be significantly improved. In the case of a single concomitant variable, we consider the ranking mechanism using the concomitant variable. In the case of multiple concomitant variables, we consider unbalanced multi-layer RSS schemes.

We consider the following general unbalanced RSS scheme. Note that the index r used in the following represents a multiple rank in a multi-layer RSS and has its usual meaning in the case of a single concomitant variable. For $r = 1, \ldots, k$, let n_r random sets of sampling units, each of size k, be drawn from the population. For each of these sets, let its units be ranked according to a certain ranking mechanism determined by $\boldsymbol{X}$. Only for the unit with rank r, is the value of Y measured. Denote the measured value of Y together with that of $\boldsymbol{X}$ on this unit by $(Y_{[r]i}, \boldsymbol{X}_{[r]i})$. The sampling scheme yields an unbalanced data set as follows:

$$(Y_{[r]i}, \boldsymbol{X}_{[r]i}) : \quad i = 1, \ldots, n_r; \; r = 1, \ldots, k. \tag{6.17}$$

Let $n = \sum_{r=1}^{k} n_r$. To allow an asymptotic analysis, assume that, as $n \to \infty$, $n_r/n \to q_r$, $r = 1, \ldots, k$. Let $\boldsymbol{q} = (q_1, \ldots, q_k)^T$ and $\boldsymbol{q}_n = (n_1, \ldots, n_k)^T/n$. We refer to $\boldsymbol{q}_n$ as an allocation vector, $\boldsymbol{q}$ as an asymptotic allocation vector.

In this section, we do not distinguish the intercept of the regression function from the other coefficients and denote all the coefficients by the vector $\boldsymbol{\beta}$. The estimation of $\boldsymbol{\beta}$ can again be done by ordinary least squares estimation. Let X_{RSS} and $\boldsymbol{y}$ be the design matrix and the response vector in the regression model with the unbalanced data. The least square estimate of $\boldsymbol{\beta}$ is given by

$$\hat{\boldsymbol{\beta}}(\boldsymbol{q}_n) = (X'_{\text{RSS}} X_{\text{RSS}})^{-1} X'_{\text{RSS}} \boldsymbol{y}.$$

The covariance matrix of $\hat{\boldsymbol{\beta}}(\boldsymbol{q}_n)$ is given by

$$\text{Var}(\hat{\boldsymbol{\beta}}(\boldsymbol{q}_n)) = \frac{\sigma_\epsilon^2}{n} E\left(\frac{X'_{\text{RSS}} X_{\text{RSS}}}{n} \right)^{-1}, \tag{6.18}$$

where the expectation is taken with respect to the distribution of the X's. Note that $\hat{\beta}(\boldsymbol{q}_n)$ is an unbiased estimate of β, regardless of the allocation vector $\boldsymbol{q}_n$. Our goal is then to find an allocation vector such that the covariance matrix of $\hat{\boldsymbol{\beta}}(\boldsymbol{q}_n)$ is the smallest in a certain sense. We take up this in the next section.

6.4.1 Asymptotically optimal criteria

The determination of exact optimal allocation vectors is not analytically tractable. Instead, we shall consider asymptotically optimal allocation vectors. We consider several types of optimality: (i) *A-optimality*, which entails the minimization of the trace of the asymptotic covariance matrix of $\hat{\boldsymbol{\beta}}(\boldsymbol{q}_n)$; (ii) *D-optimality*, which entails the minimization of the determinant of the asymptotic covariance matrix of $\hat{\boldsymbol{\beta}}(\boldsymbol{q}_n)$ and (iii) *IMSE-optimality*, which entails the minimization of the integrated mean square error (IMSE) of the estimated regression function that is a more relevant criterion when our main concern is the prediction of the response variable. The IMSE of the estimated regression function is defined as

$$\text{IMSE}(\boldsymbol{q}) = \int \text{MSE}(\hat{\boldsymbol{\beta}}' \boldsymbol{x}) f(\boldsymbol{x}) d\boldsymbol{x},$$

where $f(\boldsymbol{x})$ is the joint density function of $\boldsymbol{X}$.

The asymptotically optimal allocation vector can be obtained by minimizing the asymptotic form of the optimality criterion. In what follows, we concentrate on the special case of polynomial regression with a single concomitant variable. The case of multiple regression with multi-layer RSS can be treated in a similar way.

In the polynomial regression case, $\boldsymbol{X} = (1, X, X^2, \ldots, X^p)'$, and the ranking in RSS is done according to the magnitude of X. Let m_j and $m_{j(r)}$ denote, respectively, the jth moment of X and the jth moment of the rth order statistic of a simple random sample of size k from the distribution of X. As $n \to \infty$ such that $n_r/n \to q_r$, we have

$$\frac{1}{n}\sum_{r=1}^{k}\sum_{i=1}^{n_r} X_{(r)i}^j \to \sum_{r=1}^{k} q_r m_{j(r)}.$$

Therefore, the factor $E[n(X_{\text{RSS}}' X_{\text{RSS}})^{-1}]$ of the variance of $\hat{\boldsymbol{\beta}}(\boldsymbol{q}_n)$ converges to $V(\boldsymbol{q})$ where

$$V^{-1}(\boldsymbol{q}) = \sum_{r=1}^{k} q_r \begin{pmatrix} 1 & m_{1(r)} & \cdots & m_{p(r)} \\ m_{1(r)} & m_{2(r)} & \cdots & m_{p+1(r)} \\ \cdots & \cdots & \cdots & \cdots \\ m_{p(r)} & m_{p+1(r)} & \cdots & m_{2p(r)} \end{pmatrix}.$$

Notice that when the RSS is balanced the above matrix reduces to

$$V_0^{-1} = \begin{pmatrix} 1 & m_1 & \cdots & m_p \\ m_1 & m_2 & \cdots & m_{p+1} \\ \cdots & \cdots & \cdots & \cdots \\ m_p & m_{p+1} & \cdots & m_{2p} \end{pmatrix},$$

which corresponds to the asymptotic variance of the least square estimate of $\boldsymbol{\beta}$ under an SRS scheme. The criterion IMSE is given, in the case of polynomial regression, by

$$\text{IMSE}(\boldsymbol{q}) = \frac{\sigma_\epsilon^2}{n}\sum_{i=0}^{p}\sum_{j=0}^{p} m_{i+j} V_{ij}(\boldsymbol{q}),$$

where $V_{ij}(\boldsymbol{q})$ is the (i,j)th entry of $V(\boldsymbol{q})$ and m_{i+j} is the $(i+j)$th moment of X.

To carry out the optimization, the moments associated with the distribution of X must be supplied. In the case that the distribution of X is known, these moments can be computed numerically. In general, the moments can be estimated from the data. Since the number of measurements on the X-value is usually large, accurate estimates of the moments can be obtained. Let $\mathcal{X} = \{X_1, X_2, \ldots, X_N\}$ be the pooled simple random sample from the distribution of X. The estimates of the moments of X are given by

$$\hat{m}_j = \frac{1}{N}\sum_{i=1}^{N} X_i^j, \; j = 1, 2, \ldots$$

For the moments of the order statistics, the bootstrap procedure that follows is required for their estimation. First, draw B bootstrap samples of size k from $\mathcal{X}$ with replacement:

$$(X_{1b}^*, \cdots, X_{kb}^*),\ b = 1, \cdots, B.$$

Then order each of these samples to yield

$$(X_{(1)b}^* \leq \cdots \leq X_{(k)b}^*),\ b = 1, \cdots, B.$$

Finally, the estimates of the moments of the order statistics are approximated by

$$\hat{m}_{j(r)} = \frac{1}{B}\sum_{b=1}^{B} X_{(r)b}^{*j},\ r = 1, \cdots, k; j = 1, 2, \ldots$$

Since, as $B \to \infty$, $\frac{1}{k}\sum_{r=1}^{k} \hat{m}_{j(r)} \to \hat{m}_j$, the bootstrap size B is determined such that $\max_j |\frac{1}{k}\sum_{r=1}^{k} \hat{m}_{j(r)} - \hat{m}_j| \leq \epsilon$ for some specified precision ϵ.

6.4.2 Asymptotically optimal schemes for simple linear regression

In this section we consider asymptotically optimal schemes in more detail for simple linear regression. Denote $\sum_{r=1}^{k} q_r m_{j(r)}$ by $m_j(\boldsymbol{q})$, $j = 1, 2$. The asymptotic variances of the least square estimates of the intercept and slope, $\hat{\beta}_0(\boldsymbol{q}_n)$ and $\hat{\beta}_1(\boldsymbol{q}_n)$, and their covariance are explicitly given by

$$\sigma_0^2(\boldsymbol{q}) = \frac{\sigma_\epsilon^2}{n}\frac{m_2(\boldsymbol{q})}{m_2(\boldsymbol{q}) - [m_1(\boldsymbol{q})]^2},$$

$$\sigma_1^2(\boldsymbol{q}) = \frac{\sigma_\epsilon^2}{n}\frac{1}{m_2(\boldsymbol{q}) - [m_1(\boldsymbol{q})]^2},$$

$$\sigma_{01}(\boldsymbol{q}) = -\frac{\sigma_\epsilon^2}{n}\frac{m_1(\boldsymbol{q})}{m_2(\boldsymbol{q}) - [m_1(\boldsymbol{q})]^2}.$$

The objective functions to be minimized for the optimality criteria mentioned in the previous section can be expressed explicitly as follows:

- *A-optimality*:

$$g_A(\boldsymbol{q}) = \frac{1 + [m_1(\boldsymbol{q})]^2}{m_2(\boldsymbol{q}) - [m_1(\boldsymbol{q})]^2}.$$

- *D-optimality*:

$$g_D(\boldsymbol{q}) = \frac{1}{m_2(\boldsymbol{q}) - [m_1(\boldsymbol{q})]^2} \propto \sigma_1^2(\boldsymbol{q}).$$

- *IMSE-optimality*:

$$g_I(\boldsymbol{q}) = \frac{\sigma_X^2 + [m_1 - m_1(\boldsymbol{q})]^2}{m_2(\boldsymbol{q}) - [m_1(\boldsymbol{q})]^2}.$$

The above three objective functions can all be expressed in a unified form: $g(\boldsymbol{q}) = G(m_1(\boldsymbol{q}), m_2(\boldsymbol{q}))$, where the bivariate function $G(\cdot, \cdot)$ is positive, differentiable and decreasing in its second argument. We have the following result:

Theorem 6.2. *Let $g(\boldsymbol{q}) = G(m_1(\boldsymbol{q}), m_2(\boldsymbol{q}))$. Suppose that $\frac{\partial G}{\partial m_1} \neq 0$ and $\frac{\partial G}{\partial m_2} < 0$. Then, on the simplex $\{\boldsymbol{q} = (q_1, \ldots, q_k)' : q_r \geq 0, r = 1, \ldots, k; \sum q_r = 1\}$, the function $g(\boldsymbol{q})$ attains its minimum at a point $\boldsymbol{q}^* = (p, 0, \ldots, 0, 1-p)$.*

The proof of this Theorem is given in the appendix. The Theorem implies that, by using any of the optimality criteria listed above, the asymptotically optimal sampling scheme for the RSS only relies on the smallest and the largest rank statistics. The Theorem also has an important implication for the computation of the optimal scheme. The optimal scheme is obtained by minimizing the objective function $g(\boldsymbol{q})$ with respect to $\boldsymbol{q}$ which involves a domain of a convex hull of dimension k. The theorem implies that, to obtain the optimal scheme, only the following univariate function needs to be minimized:

$$g(p) = G(pm_{1(1)} + (1-p)m_{1(k)}, pm_{2(1)} + (1-p)m_{2(k)}).$$

In practical problems, one is either concerned with whether or not the response variable is related to the predictor variable, or concerned with the prediction of the response variable by using the regression function. The appropriate optimality to consider for the two situations are respectively the D-optimality and the IMSE-optimality. Therefore, we discuss these two optimalities in more details.

Under the criterion of D-optimality, the optimal p, the proportion corresponding to the smallest rank statistic, is explicitly given by

$$p = \frac{2m_{1(k)}[m_{1(k)} - m_{1(1)}] + m_{2(1)} - m_{2(k)}}{2[m_{1(k)} - m_{1(1)}]^2}.$$

When the distribution of X is symmetric, p reduces to $p = 1/2$ by the following symmetry properties:

$$m_{1(r)} - m_1 = m_1 - m_{1(k-r+1)}, \quad \sigma^2_{(r)} = \sigma^2_{(k-r+1)},$$

where $\sigma^2_{(r)}$ denotes the variance of the rth order statistics. Furthermore, under the symmetry assumption, the D-optimality and the IMSE-optimality are indeed equivalent. Let $\tilde{\boldsymbol{q}}$ denote the vector obtained from $\boldsymbol{q}$ by reversing the components of $\boldsymbol{q}$, i.e., $\tilde{q}_r = q_{k-r+1}, r = 1, \ldots, k$. It follows from the symmetry properties that, for any $\boldsymbol{q}$,

$$D((\boldsymbol{q} + \tilde{\boldsymbol{q}})/2) \leq D(\boldsymbol{q}) \text{ and } \mathrm{IMSE}((\boldsymbol{q} + \tilde{\boldsymbol{q}})/2) \leq \mathrm{IMSE}(\boldsymbol{q}).$$

This implies that the minimization of $D(\boldsymbol{q})$ and IMSE($\boldsymbol{q}$) can be confined on the $\boldsymbol{q}$'s such that $q_r = q_{k-r+1}$. The Theorem then implies that the optimal p equals 1/2 for both D-optimality and the IMSE-optimality. It needs to be pointed out that, in general, the two optimalities considered above are not equivalent, and the optimal proportion p is not necessarily 1/2.

In the remainder of this section, we consider the asymptotic relative efficiency of the D-optimal RSS and the IMSE-optimal RSS with respect to

Table 6.6. The asymptotic relative efficiency of the D-optimal and IMSE-optimal RSS with respect to SRS for selected X-distributions

	Uniform		Normal		Exponential		Extreme-value	
k	ARE_D	ARE_I	ARE_D	ARE_I	ARE_D	ARE_I	ARE_D	ARE_I
5	1.571	1.222	1.800	1.286	1.950	1.269	1.840	1.041
6	1.714	1.263	2.022	1.338	2.166	1.312	2.065	1.104
7	1.833	1.294	2.220	1.379	2.360	1.345	2.268	1.154
8	1.933	1.318	2.400	1.412	2.537	1.373	2.454	1.195
9	2.018	1.337	2.563	1.439	2.702	1.395	2.624	1.229
10	2.091	1.353	2.712	1.461	2.855	1.415	2.781	1.258

the balanced RSS or SRS for estimation of β_1 and prediction of the response variable respectively. With the balanced RSS or SRS, the asymptotic variance of $\hat{\beta}_1$ and the IMSE are given by

$$\sigma_1^2 = \frac{\sigma_\epsilon^2}{n\sigma_X^2} \text{ and } \text{IMSE} = \frac{2\sigma_\epsilon^2}{n}.$$

The asymptotic relative efficiency of the D-optimal RSS for the estimation of β_1 is defined as

$$\text{ARE}_D = \frac{\sigma_1^2}{\sigma_1^2(q^*)} = \frac{pm_{2(1)} + (1-p)m_{2(k)} - [pm_{1(1)} + (1-p)m_{1(k)}]^2}{\sigma_X^2}.$$

The asymptotic relative efficiency of the IMSE-optimal RSS for the prediction of the response variable is defined as

$$\begin{aligned}\text{ARE}_I &= \frac{\text{IMSE}}{\text{IMSE}(q^*)}\\ &= 2\left[1 + \frac{\sigma_X^2 + [m_1 - pm_{1(1)} - (1-p)m_{1(k)}]^2}{pm_{2(1)} + (1-p)m_{2(k)} - [pm_{1(1)} + (1-p)m_{1(k)}]^2}\right]^{-1}.\end{aligned}$$

When the distribution of X belongs to a location-scale family, the asymptotic relative efficiencies do not depend on either the location parameter or the scale parameter. To shed some light on how much efficiency can be gained by using the optimal RSS schemes, the asymptotic relative efficiencies of the D-optimal RSS for the estimation of the slope and the IMSE-optimal RSS for the prediction of the response variable are computed for set size $k = 5, \ldots, 10$ and four distributions of X: Uniform, Normal, Exponential and Extreme-value. These efficiencies are given in Table 6.6 which illustrates that the efficiency gain is tremendous.

6.4.3 Applications

(i) Aging study of Tenualosa ilisha. Biologists are often interested in the estimation of animal growth as this plays a crucial role in determining manage-

ment strategies. Growth is conventionally assumed to be a function of age. However, the determination of age is extremely difficult and costly for certain animals. This is especially the case for most fish species and crustaceans. In this subsection, we look into a study of the fish species *Tenualosa ilisha* in Bangladesh with an annual production of 200,000 metric tonnes (Blaber et al. [23]). One important part of the study on this species is to describe its growth. In this study, a sample of 1299 sub-adult fishes (length $\geq$ 50 mm) were collected between May 1996 and August 1997. The length and weight were measured on all the sampled fishes. However, the age was determined only for a sub-sample of 184 fishes due to cost consideration. In order to determine the age of a fish, one of its otoliths is removed, cleaned and sent in a plastic bag to a lab where it is embedded onto a microscope slide with thermoplastic cement and polished with wet and dry sandpaper until its mid-plane is reached, then the polished otolith is viewed under immersion oil on a video screen attached to a microscope and the daily rings are counted along the longitudinal axis towards the posterior of the otolith. A cost-effective sampling method is extremely desirable in this example.

We illustrate in the following how the method we developed in the previous sections can be used in this example to reduce cost and increase efficiency. We consider the relationship between the length (mm) and the age (in days) of subadults for this species. For the sub-sample of 184 fishes, the scatter plot of the age and length in log-scale is given in Figure 6.1. A linear model appears to be quite appropriate. The fitted regression line log(age) = 0.0469+ 1.0615 log(length) is imposed on the scatter plot in Figure 6.1. Consider the RSS with set size $k = 10$. To obtain the optimal RSS scheme, the moments are estimated using the full sample of 1299 fishes. The estimated first two moments of log(length) are: $m_1 = 5.5521, m_2 = 31.0044$. The estimated first two moments of the smallest and the largest rank statistics are: $m_{1(1)} = 4.73, m_{1(10)} = 5.98, m_{2(1)} = 22.66, m_{2(10)} = 35.80$. Based on the estimated moments, the IMSE-optimal RSS scheme is that the age is to be determined on the unit with the smallest rank for 41% of the ranked sets and on the unit with the largest rank for the other 59% of the ranked sets. The optimal scheme results in an asymptotic relative efficiency 1.44 in terms of IMSE.

A simulation study is carried out to investigate how the asymptotically optimal scheme performs in finite sample cases. In this simulation, we take the fitted regression line as if it is the true one underlining the relationship between the length and age. The age for all 1299 fishes in the full sample is generated according to the following linear model:

$$\log(\text{age}) = 0.0469 + 1.0615 \log(\text{length}) + \epsilon,$$

where ϵ follows a normal distribution with mean zero and standard deviation 0.1796 which is the square root of the MSE of the fitting based on the subsample of 184 fishes. Then simple random samples and ranked-set samples with the optimal scheme, each of size 129, are drawn from the full sample with simulated ages. Based on 4000 simulated samples under each of the two

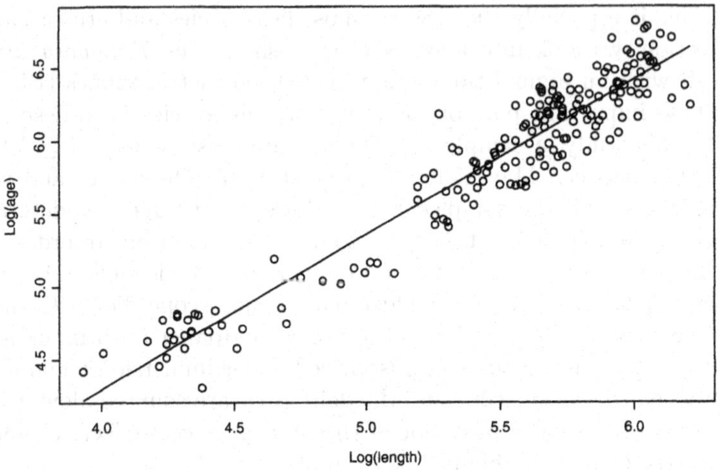

Fig. 6.1. The scatter plot of the fish aging data with fitted line superimposed.

sampling schemes, the IMSE of the two sampling schemes are estimated and the relative efficiency in terms of IMSE is approximated as 1.3879, which is quite close to the asymptotic relative efficiency. The simulated result implies that, an RSS sample of size 133 is equivalent to an SRS simple of size 184 in terms of prediction accuracy measured by IMSE. In other words, 28% of the cost for the lab investigation can be reduced.

(ii) Carcinogenic Biomarkers for Lung Cancer. Smoking is known to be a causal factor for lung cancer. However, the carcinogenic pathway from smoking to lung cancer is yet to be discovered. A case-control study was carried out in Massachusetts General Hospital between December 1992 and December 1999 to better understand the relationship among smoking exposure, carcinogenic biomarkers and disease status (Huang et al, [64])

Three bio-markers were under consideration in this study: polyphenol DNA adducts, micronuclei (MI) and sister chromatid exchanges (SCE). It is believed that DNA adduct formation is a critical step in lung cancer initiation from genotoxic agents. It is also believed to be one of the best indicators of genetic damage due to exposure to toxins or carcinogens. DNA adducts were measured on blood mononuclear cells, a long-lived component of white blood cells. MIs are chromosomal fragments or whole chromosomes excluded from the nucleus at mitosis. Their frequency is considered to be a marker of effect of smoking exposure. SCEs involve the reciprocal exchange of genetic material during cell replication. It can be induced from exposure to certain mutagens and carcinogens. MI and SCE were measured in mononuclear cells on subjects enrolled between December 1992 and April 1994. More details can be found in Cheng et al. [45] [46], Duell et al. [51].

Although these biomarkers play important roles in the pathway from exposure to disease and are often used as indicators of exposure to mutagenic and carcinogenic agents, it is not clear if they are directly associated with exposure level. For example, Cheng et al [46] did not find significant association between the mean MI frequency and smoking status among either cases or controls, although the mean MI frequency for cases was significantly higher than that for controls.

One of the purposes of the study is to find out whether or not the three bio-markers can be used as surrogates of smoking exposure. We will therefore consider how to select subjects for quantification of biomarkers in order to maximize the power of testing the association between biomarkers and smoking exposure. In this study, 1784 subjects with records of smoking exposure in pack-years were recruited, including both cases and controls. However, there were only 79, 97, and 152 subjects that have measurements on DNA adduct, MI, and SCE, respectively. More details of the marker data are described in Huang et al. [64]. Based on the data mentioned above, Huang et al. [64] found that each of the bio-markers, in standardized log scale, has somewhat a quadratic relationship with smoking exposure in units of 50 pack-years, i.e., the following relationship holds:

$$Y = \alpha + \beta_1 X + \beta_2 X^2 + \epsilon, \tag{6.19}$$

where $Y = [\log(Z) - \overline{\log(Z)}]/sd(\log(Z))$ is the standardized log of bio-marker value Z and X = pack-years/50. Estimates of $(\alpha, \beta_1, \beta_2)$ for the three bio-markers obtained by Huang et al. [64] are given below.

	α	β_1	β_2
DNA:	−0.616	0.243	−0.149
MI:	−0.324	0.300	−0.094
SCE:	−0.344	0.537	−0.138

In the following, we use the setting of the study above to illustrate how efficiency can be improved by using the optimal RSS in quantifying the relationship between the smoking exposure and the bio-markers. Like in the previous example, we take $k = 10$ and estimate the moments of X: $m_j, m_{j(r)} : j = 1, \ldots, 4; r = 1, \ldots, 10$ from the full sample of 1784 subjects. Based on the estimated moments, we seek the optimal allocation vector in the RSS by minimizing the sum of the asymptotic variances of $\hat{\beta}_1$ and $\hat{\beta}_2$. It is found that the optimal allocation vector $\boldsymbol{q}^* = (0.36, 0, \ldots, 0, 0.64)$. The asymptotic relative efficiency of the optimal RSS with respect to SRS in terms of the sum of the asymptotic variances is 1.93. Note that the optimal RSS scheme is determined by the distribution of the smoking exposure only, which is not bio-marker specific.

To illustrate how the optimal RSS performs for a fixed sample size, the following simulation study is carried out. Pseudo-values of DNA adduct, MI and

Table 6.7. The variances and relative efficiencies for the lung cancer study based on 2000 simulations

	DNA	MI	SCE
$V(\hat{\beta}_{1\text{SRS}})$	0.1656	0.1675	0.1595
$V(\hat{\beta}_{1\text{RSS}})$	0.0888	0.0738	0.0684
$\text{RE}(\beta_1)$	1.8646	2.2699	2.3304
$V(\hat{\beta}_{2\text{SRS}})$	0.03030	0.0293	0.0288
$V(\hat{\beta}_{2\text{RSS}})$	0.0099	0.0067	0.0059
$\text{RE}(\beta_2)$	3.0472	4.3473	4.8935
$\text{RE}(\beta_1, \beta_2)$	1.9836	2.4438	2.5333

SCE are generated using (6.19) with the estimated regression coefficients and i.i.d. standard normal errors. 2000 simple random samples and 2000 ranked-set samples with the optimal scheme, each of size 79, are drawn from the full sample with pseudo-values of the bio-markers; a quadratic regression function is fitted for each of the markers with each sample; and finally the variances of $\hat{\beta}_1$ and $\hat{\beta}_2$ for each marker are approximated using the estimates based on the simulated samples. The approximated variances and the relative efficiencies are given in Table 6.7. The third rows of the first and the second portion of the table give the relative efficiencies in terms of the variances of the estimates of β_1 and in terms of the variances of the estimates of β_2 respectively. The last row of the table gives the relative efficiencies in terms of the A-optimality criterion, i.e., $\text{Var}(\hat{\beta}_1) + \text{Var}(\hat{\beta}_2)$. In terms of the A-optimality, the optimal RSS is about twice as efficient as the SRS.

6.5 Technical details*

Derivation of $\text{Var}[\bar{\boldsymbol{X}}_T - \bar{\boldsymbol{X}}_{\text{RSS}}]$

In the following we derive

$$\text{Var}[\bar{\boldsymbol{X}}_T - \bar{\boldsymbol{X}}_{\text{RSS}}] = \frac{1}{mk}\Sigma_{\text{RSS}} - \frac{1}{mk^2}\Sigma. \tag{6.20}$$

It suffices to consider the univariate version of (6.20), i.e.,

$$\text{Var}[(\bar{X}_T - \bar{X}_{\text{RSS}})] = \frac{\sigma^2_{\text{RSS}}}{mk} - \frac{\sigma^2_X}{mk^2}, \tag{6.21}$$

where $\sigma^2_{\text{RSS}} = (1/k)\sum_{r=1}^k \sigma^2_{(r)}$ and $\sigma^2_X, \sigma^2_{(r)}$ are the variances of X and $X_{(r)}$ respectively. Let

$$\bar{X}_C = \frac{1}{mk(k-1)}\sum_{r=1}^{k}\sum_{i=1}^{m}[\sum_{s=1}^{k} X_{rsi} - X_{(r)i}].$$

Then we can write

$$\bar{X}_T = \frac{1}{k}\bar{X}_{\text{RSS}} + \frac{k-1}{k}\bar{X}_C, \tag{6.22}$$
$$\bar{X}_T - \bar{X}_{\text{RSS}} = \frac{k-1}{k}(\bar{X}_C - \bar{X}_{\text{RSS}}).$$

Hence,

$$\begin{aligned}&\text{Var}(\bar{X}_T - \bar{X}_{\text{RSS}})\\ =&\left(\frac{k-1}{k}\right)^2[\text{Var}(\bar{X}_{\text{RSS}})+\text{Var}(\bar{X}_C)-2\text{Cov}(\bar{X}_{\text{RSS}},\bar{X}_C)].\end{aligned} \tag{6.23}$$

It follows from (6.22) that

$$\begin{aligned}&\text{Cov}(\bar{X}_{\text{RSS}},\bar{X}_C)\\ =&\frac{1}{2(k-1)}[(1/m)\sigma_X^2-\text{Var}(\bar{X}_{\text{RSS}})-(k-1)^2\text{Var}(\bar{X}_C)].\end{aligned} \tag{6.24}$$

Substituting (6.24) into (6.23), we have

$$\begin{aligned}&\text{Var}(\bar{X}_T-\bar{X}_{\text{RSS}})\\ =&\left(\frac{k-1}{k}\right)^2\left[\frac{k}{k-1}\text{Var}(\bar{X}_{\text{RSS}})+k\text{Var}(\bar{X}_C)-\frac{\sigma_X^2}{m(k-1)}\right].\end{aligned} \tag{6.25}$$

Note that

$$\text{Var}(\bar{X}_{\text{RSS}}) = \frac{\sigma^2_{\text{RSS}}}{mk} \tag{6.26}$$

We need only derive the variance of $\bar{X}_C$. Let

$$U_{(-r)i} = \sum_{s=1}^{k} X_{rsi} - X_{(r)i}.$$

For a given r, the $U_{(-r)i}$'s are i.i.d. with the same distribution as $U_{(-r)} = \sum_{s=1}^{k} X_s - X_{(r)}$ where $(X_1, \ldots, X_k)$ is a simple random sample and $X_{(r)}$ is the rth order statistic of the sample. We have

$$\text{Var}(U_{(-r)}) = k\sigma_X^2 + \sigma_{(r)}^2 - 2\text{Cov}(\sum_{s=1}^{k} X_s, X_{(r)}).$$

Hence

$$\begin{aligned}&[mk(k-1)]^2\text{Var}(\bar{X}_C)\\ =&\sum_{r=1}^{k} m\text{Var}(U_{(-r)})\end{aligned}$$

$$= mk^2\sigma_X^2 + m\sum_{r=1}^{k}\sigma_{(r)}^2 - 2m\sum_{r=1}^{k}\text{Cov}(\sum_{s=1}^{k}X_s, X_{(r)})$$

$$= mk^2\sigma_X^2 + mk\sigma_{\text{RSS}}^2 - 2m\text{Cov}(\sum_{s=1}^{k}X_s, \sum_{r=1}^{k}X_r)$$

$$= mk(k-2)\sigma_X^2 + mk\sigma_{\text{RSS}}^2. \tag{6.27}$$

Substituting (6.26) and (6.27) into (6.25), we finally get 6.21.

The derivation above can be extended straightforwardly to the case of multiple concomitant variables.

Proof of Theorem 6.2

First, it follows from Corollary 2 in Chapter 4 that the optimal $\boldsymbol{q}^*$ has only two non-zero components. Assume the two non-zero components are q_r^* and q_s^* $(r < s)$. We are going to show that r and s must be 1 and k respectively. We only present the proof of $r = 1$. The proof of $s = k$ is similar and will be omitted.

In the following, we show that, if $r > 1$, there exists a different $\tilde{\boldsymbol{q}}$ at which the objective function is smaller, which contradicts to the fact that $\boldsymbol{q}^*$ is optimal. To this end, let $\nu = (m_{1(r+1)} - m_{1(r)})/(m_{1(r+1)} - m_{1(r-1)})$ and let $\tilde{\boldsymbol{q}}$ be the vector such that $\tilde{q}_{r-1} = \nu q_r^*$, $\tilde{q}_{r+1} = (1-\nu)q_r^*$ and $\tilde{q}_s = q_s^*$, if $r+1 < s$. In the case of $r+1 = s$, let $\tilde{q}_{r+1} = (1-\nu)q_r^* + q_s^*$. We will show that $g(\tilde{\boldsymbol{q}}) < g(\boldsymbol{q}^*)$, a contradiction.

It is easy to see that $m_1(\tilde{\boldsymbol{q}}) = m_1(\boldsymbol{q}^*)$. It suffices to show that $m_2(\tilde{\boldsymbol{q}}) - m_2(\boldsymbol{q}^*) > 0$. It can be easily verified that

$$m_2(\tilde{\boldsymbol{q}}) - m_2(\boldsymbol{q}^*) = q_r^*[\nu m_{2(r-1)} + (1-\nu)m_{2(r+1)} - m_{2(r)}]$$
$$= \frac{q_r^*}{m_{1(r+1)} - m_{1(r-1)}}\Delta,$$

where

$$\Delta = \begin{vmatrix} 1 & m_{1(r-1)} & m_{2(r-1)} \\ 1 & m_{1(r)} & m_{2(r)} \\ 1 & m_{1(r+1)} & m_{2(r+1)} \end{vmatrix}.$$

In the proof that follows, we make the dependence of the order statistics moments on the set size k explicit by writing $m_{j(r)}$ as $m_{j(r:k)}$ (for $j = 1$ and 2). The following relationship among the moments of the order statistics plays a crucial role in our argument:

$$rm_{j(r+1:k)} + (k-r)m_{j(r:k)} = km_{j(r:k-1)}. \tag{6.28}$$

By using (6.28), we can express the third row in the determinant Δ as

$$\frac{k}{r}(1, m_{1(r:k-1)}, m_{2(r:k-1)}) - \frac{k-r}{r}(1, m_{1(r:k)}, m_{2(r:k)}).$$

Therefore, Δ can be expressed as

$$\Delta = \frac{k}{r} \begin{vmatrix} 1 & m_{1(r-1:k)} & m_{2(r-1:k)} \\ 1 & m_{1(r:k)} & m_{2(r:k)} \\ 1 & m_{1(r:k-1)} & m_{2(r:k-1)} \end{vmatrix}$$

Using (6.28) again, express the second row in the determinant Δ as

$$\frac{k}{r-1}(1, m_{1(r-1:k-1)}, m_{2(r-1:k-1)}) - \frac{k-r+1}{r-1}(1, m_{1(r-1:k)}, m_{2(r-1:k)}).$$

Therefore, Δ can be further expressed as

$$\Delta = \frac{k^2}{r(r-1)} \begin{vmatrix} 1 & m_{1(r-1:k)} & m_{2(r-1:k)} \\ 1 & m_{1(r-1:k-1)} & m_{2(r-1:k-1)} \\ 1 & m_{1(r:k-1)} & m_{2(r:k-1)} \end{vmatrix}.$$

Applying (6.28) one more time to the row $(1, m_{1(r:k-1)}, m_{2(r:k-1)})$, we have

$$\Delta = \frac{k^2(k-1)}{r(r-1)^2} \begin{vmatrix} 1 & m_{1(r-1:k)} & m_{2(r-1:k)} \\ 1 & m_{1(r-1:k-1)} & m_{2(r-1:k-1)} \\ 1 & m_{1(r-1:k-2)} & m_{2(r-1:k-2)} \end{vmatrix}.$$

Let $X_{(r-1:k)}, Y_{(r-1:k-1)}, Z_{(r-1:k-2)}$ be independent $(r-1)$st order statistics of samples of sizes $k, k-1, k-2$ respectively from the same distribution F. Let F_X, F_Y and F_Z denote their respective distribution functions, e.g.,

$$dF_X(x) = \frac{k!}{(r-2)!(k-r+1)!} F^{r-2}(x)[1-F(x)]^{k-r+1} dF(x), \text{ etc.}$$

We can express Δ as

$$\begin{aligned} \Delta &= C_1 E \begin{vmatrix} 1 & X_{(r-1:k)} & X^2{}_{(r-1:k)} \\ 1 & Y_{(r-1:k-1)} & Y^2{}_{(r-1:k-1)} \\ 1 & Z_{(r-1:k-2)} & Z^2{}_{(r-1:k-2)} \end{vmatrix} \\ &= \int\int\int h(x,y,z) dF_X(x) dF_Y(y) dF_Z(z), \end{aligned}$$

where $h(x,y,z) = (y-x)(z-y)(z-x)$ and C_1 is a positive constant. Note that

$$\begin{aligned} h(x,y,z) &= h(z,x,y) = h(y,z,x) \\ &= -h(x,z,y) = -h(z,y,x) = -h(y,x,z). \end{aligned}$$

We split the triple integral into 6 distinct regions based on the ordering of x, y and z ($x < y < z$, $x < x < y$, ...). After re-expressing the integration variables, we have

$$\begin{aligned}
\Delta = C_1 \int\int\int_{x<y<z} & h(x,y,z)[dF_X(x)dF_Y(y)dF_Z(z) + dF_X(z)dF_Y(x)dF_Z(y) \\
& +dF_X(y)dF_Y(z)dF_Z(x) - dF_X(x)dF_Y(z)dF_Z(y) \\
& -dF_X(z)dF_Y(y)dF_Z(x) - dF_X(y)dF_Y(x)dF_Z(z)] \\
= C_2 \int\int\int_{x<y<z} & h(x,y,z)H(x,y,z) \\
& \{[1-F(x)]^2[1-F(y)] - [1-F(x)][1-F(y)]^2 \\
& +[1-F(z)]^2[1-F(x)] - [1-F(z)][1-F(x)]^2 \\
& +[1-F(y)]^2[1-F(z)] - [1-F(y)][1-F(z)]^2\} \\
& dF(x)dF(y)dF(z) \\
= C_2 \int\int\int_{x<y<z} & h(x,y,z)h(F(x),F(y),F(z))H(x,y,z)dF(x)dF(y)dF(z) \\
> 0, &
\end{aligned}$$

where C_2 is also a positive constant and

$$H(x,y,z) = [F(x)F(y)F(z)]^{r-2}[(1-F(x))(1-F(y))(1-F(z))]^{k-r-1}.$$

The proof is completed.

6.6 Bibliographic notes

The use of a single concomitant variable in RSS was first considered by Stokes [157]. For the estimation of the population mean of the variable of interest, Patil et al. [125] made a comparison between the RSS sample mean estimate with a concomitant variable and the regression estimate with SRS. Yu and Lam [173] considered the RSS regression estimate with one concomitant variable. Chen [38] made an extensive study on the properties of regression type estimates. The method of multi-layer RSS was developed in Chen and Shen [42]. A similar idea to the multi-layer RSS was considered by Al-Saleh and Zheng [4] for the estimation of bivariate characteristics. Adaptive RSS was developed by Chen [40]. The optimal RSS for regression analysis was studied by Chen and Wang [44]. Regression with some other RSS schemes were considered by Muttlak [100], Barreto and Barnett [17], and Öztürk [114].

7

Ranked Set Sampling as Data Reduction Tools

Contrary to the traditional problem facing statisticians that sample sizes are small, the data size in data mining is tremendously huge. It is common in data mining to deal with data sets in gigabytes or even terabytes. It is simply impossible to store a whole data set of such size in the central memory of a computer. However, certain statistical procedures, for instance, the computation of a quantile, require the whole data set to be processed at the same time in the central memory. Therefore, data reduction becomes a necessary step in dealing with huge data sets for those procedures. A desirable data reduction procedure should discard those data with low information contents and retain the data with high information contents. The RSS is essentially a data selection procedure that selects only those data which have high information contents. Therefore the notion of RSS can well be applied for data reduction.

In this chapter, we discuss techniques of data reduction using the notion of RSS. The design of a data reduction procedure depends on the purpose of study. We shall consider various data reduction procedures for different purposes. In Section 7.1, a procedure called remedian is introduced for the motivation of the procedures in the subsequent sections. In Section 7.2, the procedure for the purpose of retaining information on a single quantile is dealt with. In Section 7.3, the procedure for the purpose of retaining simultaneously information on several quantiles is treated. In Section 7.4, a general procedure for retaining information on a whole univariate population is discussed. In Section 7.5, data reduction procedures for multivariate populations are considered.

7.1 Remedian: a motivation

The remedian procedure was proposed by Rousseeuw and Bassett [138] for the computation of a central summary value of a huge data set. It was developed out of robustness and storage-saving concerns. The natural robust central summary value of a data set is its median. However, unlike the sample mean which can be computed by a one-pass updating procedure, the computation of the median requires a storage space of order $O(n)$ where n is the size of the data set. When n is huge, the computation of the median becomes un-tractable. The remedian procedure aims at reducing the storage space so that a median-like central summary value can be computed. Though the proposal of the remedian procedure was a separate development from RSS, a connection between the remedian procedure and RSS is obvious. This connection motivates the data reduction procedures using the notion of RSS. In this section, we describe briefly the remedian procedure and its connection with RSS.

Let the data size $n = k^m$, where k and m are integers. The remedian procedure with base k can be described as follows. First, the n observations of the data set are divided into k^{m-1} subsets, each of size k, and the median of each subset is computed, yielding k^{m-1} (first stage) medians. Next, the k^{m-1} medians are divided into k^{m-2} subsets, each of size k, and the median of each subset of medians (the remedian) is computed, yielding k^{m-2} (second stage) medians. The process continues until a single mth stage median is generated. A one-pass algorithm whose description was given by Rousseeuw and Bassett [138] can be used to carry out this procedure. The algorithm needs only storage space of order $O(km)$, that is, the original storage space $O(k^m)$ needed is reduced to $O(km)$. In fact, in the algorithm, only m arrays of length k are needed. The algorithm starts with filling the positions of the first array. Once the first array is full the k observations in the first array are sorted and their median is stored in the first position at the second array. Then the first array is re-filled with k new observations, the median of the new observations is stored in the second position of the second array. While this process continues, after k rounds, the second array is full, then the k observations in the second array are sorted and their median is stored in the first position of the third array. The second array is re-used to store the outputs from the first array. Once the third array is full, the k observations in the third array are sorted and their median is stored in the first position of the fourth array, and the third array is re-used to store outputs from the second array. The process continues this way until the mth array is full and then the last stage median is computed from the mth array. The algorithm with $m = 3$ is illustrated in Figure 7.1.

Consider the data set as a sample from a population. It is interesting to compare the sample median and the mth stage median produced from the remedian procedure as estimates of the population median. It was shown by Chen and Chen [31] that, like the sample median, the mth stage median is a consistent estimate of the population median and follows an asymptotic normal distribution. In particular, if m is fixed, the mth stage median has the same convergence rate as the sample median. It can be derived that the ratio of the asymptotic variance of the sample median and that of the mth stage median is given by $(2/\pi)^{m-1}$, c.f., Chen and Chen ([31], Theorem 2). This ratio can be considered as the proportion of the amount of information

```
FOR i3 = 1, k {
    FOR i2 = 1, k {
        FOR i1 = 1, k {
            READ A1[i1]
        }
        A2[i2] = MEDIAN (A1)
    }
    A3[i3] = MEDIAN (A2)
}
REMEDIAN = MEDIAN(A3)
```

Fig. 7.1. An illustrating algorithm for the remedian procedure

about the population median retained by the remedian procedure. We call this ratio as information retaining ratio (IRR). It is worthy noticing that this proportion of information corresponds to a proportion of data only of size m/k^{m-1}. For example, if $k = 20$ and $m = 5$, the proportion of information retained is 0.1642557 while the proportion of data retained is only 0.0000312.

The connection between the remedian procedure and the RSS becomes obvious if we notice that the remedian procedure at each stage can in fact be considered as an RSS procedure. The set of the jth stage medians is actually an unbalanced ranked set sample of size k^{m-j} from the $(j-1)$th stage medians. Each median is indeed obtained by first ranking the elements in the corresponding subsets and then taking the one with the middle rank. Recalling the optimal RSS schemes for the estimation of a single quantile given in Section 4.5.1, we see that the remedian procedure at each stage is in fact the optimal RSS scheme for the median. This interpretation of the remedian procedure motivates us to extend it to repeated ranked-set procedures to be discussed in subsequent sections.

7.2 Repeated ranked-set procedure for a single quantile

In this section, we develop a repeated ranked set procedure for a single quantile in the light of the idea of remedian. Before we describe the procedure, let us recall some results on the estimation of quantiles using an unbalanced RSS sample. Suppose $q_1, \ldots, q_k$ are the allocation proportions for an unbalanced RSS with set size k. For any $0 < p < 1$, let $s = \sum_{r=1}^{k} q_r B(r, k-r+1, p)$ where $B(r, s, t)$ is the cumulative distribution function of the beta distribution with parameters r and s. We have derived in Section 4.2.2 that the sth sample quantile of the unbalanced RSS sample provides a consistent estimate for the pth quantile of the population from which the RSS sample is obtained. Furthermore, we can choose the allocation proportions so that the asymptotic variance of such an estimate is minimized. For a single quantile, the optimal

RSS scheme measures only one order statistic in all the ranked sets. In the following, we denote by $r^*(p)$ the optimal rank of the order statistic for the estimation of the pth quantile.

We now describe the repeated ranked set procedure for a single quantile. To fix point, let the pth quantile, denoted by $\xi(p)$, be the one of concern. We denote the original data set (huge in size) by $D^{(0)}$. Let $r_1 = r^*(p)$ and $p_1 = B(r_1, k - r_1 + 1, p)$. The procedure starts with the first stage ranked set process as follows. The observations in $D^{(0)}$ are linearly accessed in sets of size k. For each such set, the observations in the set are ranked according to their values, the r_1th order statistic is retained and the others are discarded. All these r_1th order statistics form the set $D^{(1)}$ which is referred to as the first stage data. Then, let $r_2 = r^*(p_1)$ and $p_2 = B(r_2, k - r_2 + 1, p_1)$. The second stage ranked set process proceeds by accessing linearly the first stage data in sets of size k. Like in the first stage ranked set process, each set is ranked and the r_2th order statistic is retained and the others are discarded. All these r_2th order statistics then form the second stage data $D^{(2)}$. The procedure continues this way. In general, once accesses can be made to the $(m-1)$st stage data $D^{(m-1)}$, $r_m = r^*(p_{m-1})$ and $p_m = B(r_m, k - r_m + 1, p_{m-1})$ are computed, and the mth stage ranked set process starts. The procedure does not necessarily need to proceed to the step at which only one set of quantiles is possible to be obtained. It can stop at any stage depending on the consideration of storage space. Suppose the procedure stops at stage m, then the p_mth quantile of the mth stage data $D^{(j)}$ is computed and taken as the summary measure on the pth quantile of the original data set. The reader is reminded that the quantiles of the original data set are not computable because of the huge data set size. We call the procedure stopped at the mth stage as a m-stage repeated ranked set procedure. In the implementation of a m-stage repeated ranked set procedure, we need $m - 1$ arrays of length k and an additional array whole length is determined by the data size. An illustrating algorithm with $m = 3$ is given in Figure 7.2.

Let $F^{(m)}$ denote the distribution of the observations in the jth stage data $D^{(m)}$. Note that $F^{(m)}$ is the distribution of the r_mth order statistic of a random sample of size k from the distribution $F^{(m-1)}$. Let $\xi_m(p_m)$ denote the p_mth quantile of the distribution $F^{(m)}$. Then from the results mentioned in the first paragraph of this section and the way the p_m's are defined, $\xi(p) = \xi_1(p_1) = \cdots = \xi_m(p_m) = \cdots$. Therefore, the quantile produced from the last stage data of the repeated ranked set procedure is a consistent estimate of $\xi(p)$.

We conclude this section by a discussion on the determination of the optimal rank $r^*(p)$. For some values of p and $k = 3, 4, \ldots, 10$, the optimal $r^*(p)$ is given in Table 4.8. In general the optimal rank can be obtained, in principle, by minimizing $V(\boldsymbol{q}, p)$ given in (4.12). However, in the current context, k can be much bigger than 10. For a big k, the minimization of $V(\boldsymbol{q}, p)$ is not tractable. Table 4.8 showed some evidence that the optimal allocation vector $\boldsymbol{q}^*$ has only one non-zero component. But it is difficult to have a theoretical

```
r1 = OPT.RANK (k, p)
p1 = B (r1, k-r1+1, p)
r2 = OPT.RANK (k, p1)
p2 = B (r2, k-r2+1, p1)
r3 = OPT.RANK (k, p2)
p3 = B (r3, k-r3+1, p2)
j = 1
WHILE (EOD = 0)  {
    FOR i2 = 1, k {
        FOR i1 = 1, k {
            READ A1[i1]
            IF (END OF FILE) {
                EOD = 1
                BREAK
            }
        }
        A2[i2] = RANKSTAT (A1,r1)
    }
    A3[j] = RANKSTAT (A2,r2)
    j = j+1
}
HAT.QUANTILE = QUANTILE (A3, p3)

OPT.RANK(k,p) --- A function  which gives the optimal
                  rank for the estimation of the p-th
                  quantile in the RSS with set size k.
RANKSTAT(A,r) --- A  function  which  gives  the r-th
                  smallest component of array A.
QUANTILE(A,p) --- A  function  which  gives  the p-th
                  quantile of the components of array
                  A.
```

Fig. 7.2. An illustrating algorithm for the repeated ranked set procedure for a single quantile

proof for this. Here, we introduce other methods for the choice of the ranks used in the repeated ranked set procedure. First, recall the relative efficiency of the RSS quantile estimates with respect to SRS given in Section 2.6.3. Among the estimation of all quantiles the estimation of central quantiles are more relatively efficient and the estimation of median is most relatively efficient. This motivates us to choose r^* for a given p such that the median of the r^*th order statistic is as close to the pth quantile of concern as possible. This is equivalent to choosing r^* such that $s(p) = B(r^*, k - r^* + 1, p)$ is as close to $1/2$ as possible. Another method is motivated by the idea of importance sampling which we mentioned at the end of Section 4.7. In importance sampling, in order to estimate a quantity more accurately, data values are sampled in a

Table 7.1. The information retaining ratio at one stage of the repeated ranked set procedure for a single quantile with selected p and k.

$p \backslash k$	10	20	30	40	50	75	100
0.05	0.785	0.588	0.686	0.611	0.666	0.644	0.626
0.10	0.595	0.614	0.622	0.625	0.627	0.647	0.632
0.15	0.696	0.624	0.656	0.630	0.648	0.645	0.634
0.20	0.622	0.629	0.632	0.633	0.634	0.635	0.635
0.25	0.678	0.632	0.650	0.635	0.645	0.640	0.636
0.30	0.633	0.635	0.635	0.636	0.636	0.642	0.636
0.35	0.672	0.637	0.648	0.637	0.644	0.641	0.637
0.40	0.639	0.638	0.638	0.637	0.637	0.637	0.637
0.45	0.669	0.639	0.647	0.638	0.643	0.640	0.637
0.50	0.645	0.641	0.639	0.639	0.638	0.641	0.637

way which makes it more likely for a statistic to assume a value in the vicinity of the quantity of interest. In the light of the idea of importance sampling, it is reasonable to choose r^* such that the r^*th order statistic is as close to the pth quantile of the ranked set as possible, i.e., choose $r^* = \text{Round}(kp)$, where $\text{Round}(kp)$ is the closest integer to kp. It is quite interesting to notice that when k is large the two methods we just discussed produce almost the same r^* for any p and that, for $k = 3, 4, \ldots, 10$, the r^* obtained by using the first method coincide with the optimal ranks given in Table 4.8 except only a few cases where r^* is not exactly the optimal rank but differ from the optimal rank only by 1. As a general guideline, we suggest to use the first method if k is not too large and use the second method if k is large. It is worthy remarking that the repeated ranked set procedure for a single quantile retains about the same proportion of information for the quantile as the remedian procedure does for the median. It can be checked that, for any p, the information retaining ratio after one stage of the repeated ranked set procedure is around $2/\pi$. For a demonstration, the information retaining ratio is computed for $p = 0.05, 0.10, 0.15, \ldots, 0.5$ and $k = 10, 20, 30, 40, 50, 75$ and 100 using the following formula:

$$\text{IRR}(k, p) = \frac{p(1-p)/k}{c_{r^*}(p)[1 - c_{r^*}(p)]/[d_{r^*}(p)]^2}. \tag{7.1}$$

The results are given in Table 7.1. It can be seen that the $\text{IRR}(k, p)$ does not depend on p and k too much.

7.3 Repeated ranked-set procedure for multiple quantiles

First, let us consider the following scheme. Denote by $\boldsymbol{q}^{[i]}$, $i = 1, 2, \cdots$, a sequence of allocation vectors, i.e, $\boldsymbol{q}^{[i]} = (q_1^{[i]}, \ldots, q_k^{[i]})^T$ with $0 \le q_r^{[i]}$, $\sum_{r=1}^k q_r^{[i]} = 1$.

Starting from a distribution function $F^{[0]}$ and l probabilities $p_j^{[0]}, j = 1, \ldots, l$, form the mixture distribution $F^{[1]}(x) = \sum_{r=1}^k q_r^{[1]} F_{(r)}^{[0]}(x)$, where $F_{(r)}^{[0]}$ denotes the distribution function of the rth order statistic of a sample of size k from $F^{[0]}$, and compute $p_j^{[1]} = \sum_{r=1}^k q_r^{[1]} B(r, k-r+1, p_j^{[0]})$. Then the $p_j^{[1]}$th quantile of $F^{[1]}$ is the $p_j^{[0]}$th quantile of $F^{[0]}$, $j = 1, \ldots, l$. Then from $F^{[1]}$ and $p_j^{[1]}, j = 1, \ldots, l$, form the mixture distribution $F^{[2]}(x) = \sum_{r=1}^k q_r^{[2]} F_{(r)}^{[1]}(x)$, where $F_{(r)}^{[1]}$ denotes the distribution function of the rth order statistic of a sample of size k from $F^{[1]}$, and compute $p_j^{[2]} = \sum_{r=1}^k q_r^{[2]} B(r, k-r+1, p_j^{[1]})$. Then the $p_j^{[2]}$th quantile of $F^{[2]}$ is the $p_j^{[1]}$th quantile of $F^{[1]}$ and, in turn, is the $p_j^{[0]}$th quantile of $F^{[0]}$, $j = 1, \ldots, l$. In general, once $F^{[i-1]}$ and $p_j^{[i-1]}, j = 1, \ldots, l$, are obtained, form the mixture distribution $F^{[i]}(x) = \sum_{r=1}^k q_r^{[i]} F_{(r)}^{[i-1]}(x)$, where $F_{(r)}^{[i-1]}$ denotes the distribution function of the rth order statistic of a sample of size k from $F^{[i-1]}$, and compute $p_j^{[i]} = \sum_{r=1}^k q_r^{[i]} B(r, k-r+1, p_j^{[i-1]})$. Then the $p_j^{[i]}$th quantile of $F^{[i]}$ is the $p_j^{[0]}$th quantile of $F^{[0]}$, $j = 1, \ldots, l$. Suppose the procedure stops at the mth stage. If we can generate a sample from $F^{[m]}$, the $p_j^{[m]}$th sample quantile of this sample carries the information about the $p_j^{[0]}$th quantile of $F^{[0]}$, in fact, it provides a consistent estimate of the $p_j^{[m]}$th quantile of $F^{[m]}$ and hence of the $p_j^{[0]}$th quantile of $F^{[0]}$.

A m-stage repeated ranked set procedure for multiple quantiles is indeed a procedure to generate a reduced sample of $F^{[m]}$ from the huge original data considered as a sample from a distribution, say, $F^{[0]}$. Since our goal is to reduce data but at the same time retain the desired information as much as possible, the allocation vector $\boldsymbol{q}^{[i]}$ is chosen so that the ith stage distribution $F^{[i]}$ contains as much desired information about $F^{[i-1]}$ as possible, for $i = 1, 2, \ldots$, in the sense that the $p_j^{[i]}$th quantile of a sample from $F^{[i]}$, as an estimator of the $p_j^{[i-1]}$th quantile of $F^{[i-1]}$ has the smallest variance. In principle, such an allocation vector can be determined by minimizing either the determinant or trace of the asymptotic variance-covariance matrix of the sample quantiles of the sample from $F^{[i]}$ as the estimator of the corresponding quantile of $F^{[i-1]}$, c.f., Section 4.5.2. However, for k moderate or large, which is the case in the current context, the minimization of the determinant or trace of the asymptotic variance-covariance matrix is not computationally tractable. Motivated by the single quantile case treated in the last section, we adopt the principle of importance sampling in the repeated ranked set procedures for multiple quantiles. That is, we choose the allocation vectors such that they put all the mass on the concerned quantiles only. The weights among the concerned quantiles can be determined by the importance associated with each of the quantiles. If the desired quantiles are considered equally important then equal weights are put on the quantiles. For example, if what are of concern are the first and third quartile plus the 0.05th and 0.95th quantile, an equal allocation proportion 1/4 is associated with each of the four ranks: $r_1 = \text{Round}(0.05k), r_2 = \text{Round}(0.25k), r_3 = \text{Round}(0.75k)$ and $r_4 = \text{Round}(0.95k)$.

We now describe the repeated ranked set procedure for multiple quantiles. Suppose what are of concern are the l quantiles $\xi(p_j)$, $j = 1, \dots, l$. Let $r_j^{[1]} = \text{Round}(kp_j), j = 1, \dots, l$, and $p_j^{[1]} = (1/l)\sum_{j=1}^{l} B(r_j^{[1]}, k - r_j^{[1]} + 1, p_j)$. The first stage of the procedure proceed as follows. The observations in $D^{(0)}$, the original data set, are linearly accessed in sets of size k. For each such set, the observations in the set are ranked according to their values, then, for each $j = 1, \dots, l$, the rank $r_j^{[1]}$ is chosen with probability $1/l$, and the order statistic with the chosen rank is retained and the others are discarded. This process continues until all the observations in $D^{(0)}$ are exhausted. All the retained order statistics form the first stage data set $D^{(1)}$, which is in fact a sample from the mixture distribution $F^{[1]}(x) = (1/l)\sum_{j=1}^{l} F^{[0]}_{(r_j)}(x)$. In the second stage of the procedure, let $r_j^{[2]} = \text{Round}(kp_j^{[1]})$ and compute $p_j^{[2]} = (1/l)\sum_{j=1}^{l} B(r_j^{[2]}, k - r_j^{[2]} + 1, p_j^{[1]})$ for $j = 1, \dots, l$. Then repeat the process of the first stage with $D^{(0)}$ and $r_j^{[1]}$ being replaced by $D^{(1)}$ and $r_j^{[2]}$ respectively. At the end of the second stage, the first stage data set $D^{(1)}$ is reduced to the second stage data set $D^{(2)}$, which constitutes a sample from the mixture distribution $F^{[2]}(x) = (1/l)\sum_{j=1}^{l} F^{[1]}_{(r_j^{[1]})}(x)$. With the continually reduced data sets and updated ranks, the process repeats until it stops as a certain desired stage, say, the mth stage. The last stage data set then constitutes a sample from the mixture distribution $F^{[m]}(x) = (1/l)\sum_{j=1}^{l} F^{[m-1]}_{(r_j^{[m-1]})}(x)$, and the $p_j^{[m]}$th quantile of this sample is taken as the summary statistic for $\xi(p_j)$, $j = 1, \dots, l$.

In the implementation of the repeated ranked set procedure for multiple quantiles, the same storage spaces as in the procedure for a single quantile are needed. An illustrating algorithm with $m = 3$ and $l = 3$ is given in Figure 7.3.

In the remainder of this section, we take a look at the information retaining ratio of the repeated ranked set procedure for multiple quantiles. We consider the cases of two and three quantiles. For $k = 30$ and 70 and some pairs and triplets of quantiles, the marginal information retaining ratios as defined in (7.1) are computed. The minimum, average and maximum of the marginal information retaining ratios for the pairs are presented in Table 7.2, and those for the triplets are presented in Table 7.3. Some general features manifest themselves. (i) If the quantiles of concern are far apart, the marginal information retaining ratio for l quantiles is roughly the information retaining ratio for a single quantile divided by l. This can be explained by the fact that the information about a particular quantile is mainly contained in its corresponding order statistics and the order statistics far away from the quantile

```
P = (p1,p2,p3)
R1 = ROUND (k*P)
P1 = B (R1, k-R1+1, P1)
R2 = ROUND (k*P1)
P2 = B (R2, k-R2+1, P1)
R3 = ROUND (k*P2)
P3 = B (R3, k-R3+1, P2)
j = 1
WHILE (EOD = 0)  {
    FOR i2 = 1, k {
        FOR i1 = 1, k {
            READ A1[i1]
            IF (END OF FILE) {
                EOD = 1
                BREAK
            }
        }
        RDM.NUMBER = SAMPLE (c(1,2,3))
        A2[i2] = RANKSTAT (A1,R1[RDM.NUMBER])
    }
    RDM.NUMBER = SAMPLE (c(1,2,3))
    A3[j] = RANKSTAT (A2,R2[RDM.NUMBER])
    j = j+1
}
HAT.QUANTILE = QUANTILE (A3, P3)

SAMPLE (x) --- Produces a component of x with equal
               probability.
```

Fig. 7.3. An illustrating algorithm for the repeated ranked set procedure for multiple quantiles

contain little information about it. As a consequence, for a particular quantile, only $1/l$ of the reduced data contain its information. (ii) If the quantiles of concern are close to each other, the marginal information retaining ratio is larger than the case when they are far apart. The marginal information retaining ratio increases as the closeness of the quantiles increases. This is because that the order statistics corresponding to one quantile also contain information about the other quantiles when they are close to each other. (iii) The marginal information retaining ratio seems to depend only on the closeness of the quantiles but not their particular positions. This can be attributed to the fact that the choice of the ranks of the order statistics have essentially turned the quantiles of concern to the medians of the distributions of their corresponding order statistics. (iv) The information retaining ratio does not depend on the set size k too much. As the set size increases there is only a slight decrease of the information retaining ratio. (However, it must be kept

Table 7.2. IRR at one stage of the repeated ranked set procedure for two quantiles

	$k = 30$			$k = 70$		
p	Min	Average	Max	Min	Average	Max
(0.1 0.2)	0.400	0.438	0.477	0.339	0.343	0.347
(0.2 0.3)	0.450	0.490	0.530	0.371	0.384	0.397
(0.3 0.4)	0.476	0.513	0.549	0.392	0.409	0.425
(0.4 0.5)	0.489	0.522	0.555	0.402	0.420	0.437
(0.1 0.9)	0.311	0.311	0.311	0.315	0.315	0.315
(0.2 0.8)	0.316	0.316	0.316	0.317	0.317	0.317
(0.3 0.7)	0.318	0.318	0.318	0.318	0.318	0.318
(0.4 0.6)	0.346	0.346	0.346	0.320	0.320	0.320

Table 7.3. IRR at one stage of the repeated ranked set procedure for three quantiles

	$k = 30$			$k = 70$		
p	Min	Average	Max	Min	Average	Max
(0.1 0.2 0.3)	0.305	0.347	0.411	0.231	0.253	0.280
(0.2 0.3 0.4)	0.330	0.387	0.455	0.262	0.282	0.320
(0.3 0.4 0.5)	0.343	0.406	0.476	0.269	0.298	0.342
(0.1 0.5 0.9)	0.207	0.209	0.213	0.210	0.211	0.213
(0.2 0.5 0.8)	0.211	0.212	0.214	0.211	0.212	0.213
(0.3 0.5 0.7)	0.227	0.240	0.247	0.213	0.213	0.214

in mind that, as k increases, more data reduction is done, in fact, the data is reduced at the rate $1/k$.)

7.4 Balanced repeated ranked-set procedure

The repeated ranked set procedures described in the previous sections are designed for retaining certain specific features of the original data. In this section, we consider a balanced repeated ranked set procedure for general purposes. The procedure has been studied by Al-Saleh and Al-Omari [2] in the context of sampling. However, it is more appropriate for data reduction than for sampling because of the difficulty in ranking in the context of sampling. We briefly describe the procedure first and then discuss some of its properties.

The balanced repeated ranked set procedure is similar to the procedures considered in the previous sections. The balanced repeated ranked set procedure differs only by that in each stage of the procedure the order statistics with ranks from 1 to k are retained a equal number of times while in the previous procedures only certain order statistics with pre-specified ranks are retained. In other words, in each stage of the procedure, a balanced ranked set sampling is carried out to reduce the data obtained from the previous stage to a balanced ranked set sample of that data. Specifically, let $D^{(j)}$ be the array

```
j = 1
WHILE (EOD = 0)  {
    FOR i2 = 1, k {
        FOR i1 = 1, k {
            READ A1[i1]
            IF (END OF FILE) {
                EOD = 1
                BREAK
            }
        }
        A2[i2] = RANKSTAT (A1,i2)
    }
    r = MOD (j,k)
    if (r = 0) r = k
    A3[j] = RANKSTAT (A2,r)
    j = j+1
}
```

Fig. 7.4. An illustrating algorithm for the balanced repeated ranked set procedure

of the data obtained in the jth stage. Then the $(j+1)$st stage is carried out as follows. The array $D^{(j)}$ is linearly accessed in sets of size k. For the first set, the observations in the set are ranked and the smallest order statistic is retained and the others are discarded. For the second set, the observations in the set are ranked and the second smallest order statistic is retained and the others are discarded. The process continues until the largest order statistic in the kth set is retained. Then the whole cycle of this process is repeated. The retained order statistics are stored in a new array, say $D^{(j+1)}$, in a linear order. For example, the order statistic with rank 1 in the first cycle is stored in the first position of $D^{(j+1)}$, the order statistic with rank 2 in the first cycle is stored in the second position of $D^{(j+1)}$, and so on. For the procedure with 2 stages, an illustrating algorithm is given in Figure 7.4.

Let the order statistics obtained at the jth stage be denoted by

$$X_{(r)i}^{[j]} : r = 1, \ldots, k, i = 1, 2, \ldots$$

Denote the distribution function of $X_{(r)i}^{[j]}$ by $F_{(r)}^{[j]}(x)$. Note that $X_{(r)i}^{[j]}$ is the rth smallest value of $X_{(1)i}^{[j-1]}, \ldots, X_{(k)i}^{[j-1]}$. Al-Saleh and Al-Omari [2] derived the following properties:

(i) For any j,

$$\frac{1}{k}\sum_{r=1}^{k} F_{(r)}^{[j]}(x) = F(x),$$

where $F(x)$ is the distribution function of the original data.

(ii) As $j \to \infty$, $F_{(r)}^{[j]}(x)$ converges to a distribution function given by

$$F_{(r)}^{[\infty]}(x) = \begin{cases} 0, & x < \xi_{(r-1)/k}, \\ kF(x) - (r-1), & \xi_{(r-1)/k} \le x < \xi_{r/k}, \\ 1, & x \ge \xi_{r/k}, \end{cases}$$

where ξ_p denotes the pth quantile of F, for $r = 1, \ldots, k$.

The above results have important implications. First, it is implied that the balanced repeated ranked set procedure retains the overall structure of the original data. In certain sense, the distribution of the original data can be reconstructed from the reduced data. For instance, the empirical distribution function of the reduced data can be taken as an estimate (indeed a good estimate) of the original distribution. Any specific feature of the original distribution can be estimated by the corresponding feature of the empirical distribution of the reduced data. Second, the procedure approximately stratifies the original data so that an equal number of observations are retained from the portions of the original distribution with equal probability mass. As a consequence, the final reduced data provides a much more reliable representative of the original data than a simple random sample, since a simple random sample (even of large size) could be very erratic, especially, locally.

7.5 Procedure for multivariate populations

In chapter 6, we introduced the multi-layer ranked set sampling for the inference of the variable of interest when multiple concomitant variables are available. The multi-layer ranked set sampling can be adapted to develop a data reduction procedure for multivariate populations. However, some points of view must be changed. The variables used in the ranking mechanism are no longer treated as concomitant variables. It is the joint distribution of the multiple variables that is of our concern rather than the distribution of a single response variable. We can repeat the multi-layer ranked set sampling scheme in an appropriate way. The resultant procedure will be called a repeated multi-layer ranked set procedure. In this section, we describe the repeated multi-layer ranked set procedure and establish a result similar to property (i) of the balanced repeated ranked set procedure given in the last section.

For the simplicity of notation, we describe the repeated multi-layer ranked set procedure for the bivariate case, i.e., the original data is a collection of two-dimensional vectors. Denote the original data set by $D^{(0)} = \{\boldsymbol{X}_i : i = 1, \ldots, N\}$ where $\boldsymbol{X}_i = (X_i^{[1]}, X_i^{[2]})$ and N is the size of the data set. As in the repeated ranked set procedures described in the previous sections, the repeated multi-layer ranked set procedure is carried out conceptually in stages. At each stage, there are two steps: the first step involves ranking the vectors according to their first variable and the second step involves ranking the vectors according to their second variable. For convenience, we refer to the rank arising from ranking the first variable as X_1-rank and that arising from ranking the second variable as X_2-rank. For the first stage, in step 1, the data in $D^{(0)}$ are accessed in batches of size k, the vectors in each batch are ranked according to their first variable, and only one vector in each batch with a pre-specified X_1-rank is retained. The pre-specified X_1-ranks are in cycles of length k^2 with each cycle consisting of k 1's followed by k 2's and so on. This step produces a sequence of vectors of the following form:

$$\begin{array}{cccc}
\boldsymbol{X}_{[1]1} & \boldsymbol{X}_{[1]2} & \cdots & \boldsymbol{X}_{[1]k} \\
\boldsymbol{X}_{[2]1} & \boldsymbol{X}_{[2]2} & \cdots & \boldsymbol{X}_{[2]k} \\
\cdots & \cdots & \cdots & \cdots \\
\boldsymbol{X}_{[k]1} & \boldsymbol{X}_{[k]2} & \cdots & \boldsymbol{X}_{[k]k} \\
\boldsymbol{X}_{[1]k+1} & \boldsymbol{X}_{[1]k+2} & \cdots & \boldsymbol{X}_{[1]2k} \\
\cdots & \cdots & \cdots & \cdots
\end{array} \tag{7.2}$$

where the numbers in brackets are the X_1-ranks of the vectors. In step 2, the data produced from step 1 are processed in batches of size k, that is, the vectors represented in (7.2) are processed row by row, the vectors in each row are ranked according to their values of the second variable, and only one vector in each batch with a pre-specified X_2-rank is retained. The pre-specified X_2-ranks are in the same cycles as those pre-specified X_1-ranks. This step produces a sequence of vectors as follows:

$$\begin{array}{cccc}
\boldsymbol{X}_{[1][1]1} & \boldsymbol{X}_{[2][1]1} & \cdots & \boldsymbol{X}_{[k][1]1} \\
\boldsymbol{X}_{[1][2]1} & \boldsymbol{X}_{[2][2]1} & \cdots & \boldsymbol{X}_{[k][2]1} \\
\cdots & \cdots & \cdots & \cdots \\
\boldsymbol{X}_{[1][k]1} & \boldsymbol{X}_{[2][k]1} & \cdots & \boldsymbol{X}_{[k][k]1} \\
\boldsymbol{X}_{[1][1]2} & \boldsymbol{X}_{[2][1]2} & \cdots & \boldsymbol{X}_{[k][1]2} \\
\cdots & \cdots & \cdots & \cdots
\end{array} \tag{7.3}$$

where the numbers in the first and second bracket are the X_1-ranks and X_2-ranks of the vectors respectively. At each of the subsequent stages, the process just described is repeated with the data set $D^{(0)}$ being replaced by the sequence produced from the previous stage.

In the implementation of the repeated multi-layer ranked set procedure, two arrays of length k are needed for processing the data at each stage. If the procedure consists of m stages, $2m$ arrays of length k and an additional array of unspecified length are needed. The last array stores the final reduced data. In general, for a p-variate population, each stage of the repeated multi-layer ranked set procedure consists of p steps and a total of $pm+1$ arrays are needed for processing the data. The storage space needed for the procedure is pmk plus the length of the final array. An illustrating algorithm for the procedure with 2 stages is given in Figure 7.5

Let $\boldsymbol{X}^{(j)}_{[r][s]}$ denote the data with X_1-rank r and X_2-rank s obtained in the jth stage of the procedure. Denote by $F^{(j)}_{[r][s]}(x_1, x_2)$ its corresponding distribution function. In the following, we show that

```
        j = 1
        WHILE (EOD = 0)  {
          r = MOD (j,k)
          if (r = 0) r = k
          FOR j2 = 1, k {
            FOR j1 = 1, k {
              FOR i2 = 1, k {
                FOR i1 = 1, k {
                  READ A1[i1, ]
                  IF (END OF FILE) {
                    EOD = 1
                    BREAK
                  }
                }
                A2[i1, ] = DRANKSTAT (A1,j1,1)
              }
              B1[j1, ] = DRANKSTAT (A2,j2,2)
            }
            B2[j2, ] = DRANKSTAT (B1,r,1)
          }
          RD[j, ] = DRANKSTAT (B2,r,2)
          j = j+1
        }

DRANKSTAT (A,r,i) --- Gives the row of A with rank r
                      when the rows are ordered according
                      to the i-th column of A, i = 1, 2.
MOD (j,k)    ---      Produces the remainder when j is di-
                      vided by k.
```

Fig. 7.5. An illustrating algorithm for the repeated multi-layer ranked set procedure

$$\frac{1}{k^2}\sum_{r=1}^{k}\sum_{s=1}^{k} F_{[r][s]}^{(j)}(x_1, x_2) = \frac{1}{k^2}\sum_{r=1}^{k}\sum_{s=1}^{k} F_{[r][s]}^{(j-1)}(x_1, x_2) = F(x_1, x_2), \qquad (7.4)$$

for $j = 1, 2, \ldots$, where $F(x_1, x_2)$ is the joint distribution function of the original data.

The above result follows from the following lemma.

Lemma 7.1. *Let* $\{\boldsymbol{X}_{I_1}, \boldsymbol{X}_{I_2}, \ldots, \boldsymbol{X}_{I_n}\}$ *be a collection of random variables. Let* $(R_1, R_2, \ldots, R_n)$ *be a random permutation of* $(I_1, I_2, \ldots, I_n)$*. If* R_i *is the image of* I_j *under the random permutation, we denote* $R_i = P(I_j)$*. For* $i = 1, \ldots, n$*, define* $\boldsymbol{Y}_{R_i} = \boldsymbol{X}_{I_j}$ *if* $R_i = P(I_j)$*. Let the distribution functions of* $\boldsymbol{Y}_{R_i}$ *and* $\boldsymbol{X}_{I_j}$ *be denoted by* F_{R_i} *and* F_{I_j} *respectively. Then we have*

$$\sum_{i=1}^{n} F_{R_i}(\boldsymbol{x}) = \sum_{j=1}^{n} F_{I_j}(\boldsymbol{x}).$$

PROOF:

$$
\begin{aligned}
F_{R_i}(\boldsymbol{x}) &= P(\boldsymbol{Y}_{R_i} \le \boldsymbol{x}) \\
&= \sum_{j=1}^{n} P(\boldsymbol{Y}_{R_i} \le \boldsymbol{x}, R_i = P(I_j)) \\
&= \sum_{j=1}^{n} E[I\{\boldsymbol{X}_{I_j} \le \boldsymbol{x}\} I\{R_i = P(I_j)\}],
\end{aligned}
$$

where $I\{\cdot\}$ is the usual indicator function and the binary operator "$\le$" is applied component-wise. Therefore

$$
\begin{aligned}
\sum_{i=1}^{n} F_{R_i}(\boldsymbol{x}) &= \sum_{i=1}^{n} \sum_{j=1}^{n} E[I\{\boldsymbol{X}_{I_j} \le \boldsymbol{x}\} I\{R_i = P(I_j)\}] \\
&= \sum_{j=1}^{n} E[I\{\boldsymbol{X}_{I_j} \le \boldsymbol{x}\} \sum_{i=1}^{n} I\{R_i = P(I_j)\}] \\
&= \sum_{j=1}^{n} E[I\{\boldsymbol{X}_{I_j} \le \boldsymbol{x}\} \\
&= \sum_{j=1}^{n} F_{I_j}(\boldsymbol{x}),
\end{aligned}
$$

since $\sum_{i=1}^{n} I\{R_i = P(I_j)\} = 1$. The lemma is hence proved.

Let $X_1, \ldots, X_n$ be i.i.d. random variables and $X_{(1)}, \ldots, X_{(n)}$ be their order statistics. It can be considered as a special case of the above lemma that the sum of the distribution functions of the order statistics equals to n times the common distribution function of the X_i's. However, the above lemma does not confine to the i.i.d. case. In fact, no assumption about the relationship among the $\boldsymbol{X}_{I_j}$'s is made at all. Now let k^2 2-dimensional random vectors be indexed as $\boldsymbol{X}_{rs} : r = 1, \ldots, k, s = 1, \ldots, k$. Let these vectors be "ranked" according to the scheme of the multi-layer ranked set procedure to produce "ranks" $[r][s]$. Then this scheme determines a random permutation of the original indices. Thus the above lemma applies. Since only the marginal distribution functions of the $\boldsymbol{X}_{[r][s]}$'s are of concern, no matter whether they are obtained from a common sample or from different samples that have the same joint distribution, the sum of the marginal distribution functions is the same. This verifies the result in (7.4). This result ensures that the overall structure of the original data is retained by the reduced data.

8

Case Studies

In this chapter we discuss a few case studies which clearly indicate the advantages of using a ranked set sample over those of a simple random sample. The applications of RSS considered here cover a wide range of topics from plantations of cinchona in India (Sengupta et al. [145] and Stokes [159]), to cost effective gasoline sampling (Nussbaum and Sinha [111]), browse and herbage in a pine-hardwood forest of east Texas, USA (Halls and Dell [58]), shrub phytomass in forest stands (Martin et al. [95]), single family homes sales data (Bowerman and O'Connell [29]), and tree data (Platt et al. [133]). The objective in most of these studies is not so much in a direct application of RSS to efficiently solve an estimation problem but rather to strongly support the fact that use of RSS over SRS does provide substantial advantages for the estimation of both mean and variance. Thus in most of these studies samples from the entire population are made available so that a comparison can be made among the population parameters and their estimates which are obtained via SRS and RSS. Moreover, in case of unbalanced RSS, since the population values are all available, it is indeed possible then to use the optimum choice of the *replications* of different order statistics. The two latest applications appear in Sarikavanij et al. [144].

8.1 Plantations of cinchona

This problem is reported in Stokes [159]. Cinchona plants from which quinine (medical cure from malaria) is extracted were grown on steep hills in southern India during 1940's. Sengupta et al. [145] undertook a study to estimate yield of dry bark and quinine contents from these plants. Measurement of dry bark yield typically requires uprooting the plant, stripping the bark, and drying it until the weight settles down to a steady value. It is thus clear that an actual measurement in this case is rather time consuming and perhaps costly too, indicating that an application of RSS would be worthwhile to pursue.

Table 8.1. Ranked set sample of volumes of bark plants and corresponding dry bark weights (in ounces)

r	i = 1	2	3
1	(185.53,11.33)	(187.33,12.41)	(183.60,10.18)
2	(192.06,15.25)	(191.57,14.95)	(198.71,19.23)
3	(190.14,14.10)	(198.19,18.92)	(196.33,17.80)
4	(204.55,22.73)	(205.33,23.19)	(210.42,26.24)
5	(221.57,32.92)	(217.07,30.22)	(207.67,24.59)

r	i = 4	5	6
1	(186.72,12.04)	(179.80,7.91)	(185.89,11.55)
2	(191.62,14.98)	(187.05,12.25)	(201.65,20.99)
3	(198.51,19.11)	(211.14,26.67)	(214.84,28.89)
4	(199.30,19.58)	(195.53,17.32)	(203.36,22.01)
5	(217.35,30.39)	(210.30,26.17)	(223.25,33.92)

Fortunately, in this context a simultaneous investigation showed that dry bark yield was very highly correlated with the volume of bark per plant which can be computed rather easily from the height of the plant and its girths and thicknesses at several heights, without destroying the plant. We thus have a handy cheap covariate (X: volume of bark plant) which can be readily used to rank the primary variable (Y: dry bark weight).

Unfortunately, the raw data from the study by Sengupta et al. [145] are not available, and Stokes [159] used simulation with $\rho = 0.9$, $\mu_y = 20.08$ and $\sigma_y = 7.79$, resembling the corresponding estimated values in the study by Sengupta et al.[145]. This seems to be an incomplete information because we also need to know the values of μ_x and σ_x in order to generate bivariate samples. We have taken these values as $\mu_x = 200$ and $\sigma_x = 10$. We generated 150 bivariate normal samples of (X, Y) under the population parameter values: $\mu_y = 20$, $\sigma_y = 8$, $\mu_x = 200$, $\sigma_x = 10$, and $\rho = 0.9$. We then divided the data into 6 sets of 25 observations each, 25 observations within each set being in turn divided into 5×5 as a square. This was followed by a ranked set sample of X-values, and we then measured the corresponding judgment ordered Y-values. These values $\{x_{(r)i}, y_{[r]i}\}$ of the ranked set sample of (X, Y) for $k = 5$ and $n = 6$ are given below in Table 8.1.

We can easily compute the RSS-based estimate of μ_y and the various estimates of σ_y^2 from this data on the basis of our discussion in Chapter 2, Section 2.4. These are given below in Table 8.2 along with an estimate of the variance of $\hat{\mu}_{yrss}$ on the basis of the formula: $\hat{var}(\hat{\mu}_{yrss}) = \sum_{r=1}^{k} \hat{\sigma}_{(r)}^2 / nk^2$. Obviously $\hat{\sigma}_{(r)}^2$'s are obtained from the 5 rows of Table 8.1. It is readily observed that these values are much smaller than $\hat{\sigma}_y^2/30 = 2.77$, which is the estimated variance of a SRS mean based on a y-sample of size 30.

Table 8.2. Estimates of population mean and variance of dry bark weights (in ounces)

Method	Mean(y)	Variance(y)	Variance of mean(y)
BRSS	19.93	$\hat{\sigma}^2_{Stokes}$=52.07	0.45
		$\hat{\sigma}^2_1$=50.91	
		$\hat{\sigma}^2_2$=50.79	
		$\hat{\sigma}^2_3$=50.21	
		$\hat{\sigma}^2_4$=53.79	
REG	20	$\hat{\sigma}^2_{Stokes}$=52.07	(based on $\hat{\sigma}^2_{Stokes}$) 1.35×10^{-9}
		$\hat{\sigma}^2_1$=50.91	(based on $\hat{\sigma}^2_1$) 1.32×10^{-9}
		$\hat{\sigma}^2_2$=50.79	(based on $\hat{\sigma}^2_2$) 1.32×10^{-9}
		$\hat{\sigma}^2_3$=50.21	(based on $\hat{\sigma}^2_3$) 1.30×10^{-9}
		$\hat{\sigma}^2_4$=53.79	(based on $\hat{\sigma}^2_4$) 1.40×10^{-9}

Table 8.3. Summary statistics of estimates of population mean and variance for 1000 RSS samples

variable	Mean	S.D.	Median
$\hat{\mu}_{RSS}$	19.99	0.88	19.98
$\hat{var}(\hat{\mu}_{RSS})$	0.78	0.22	0.75
$\hat{\sigma}^2_{Stokes}$	65.67	14.93	64.55
$\hat{\sigma}^2_1$	64.27	14.53	63.21
$\hat{\sigma}^2_2$	64.26	14.54	63.16
$\hat{\sigma}^2_3$	64.23	14.81	62.74
$\hat{\sigma}^2_4$	64.42	17.90	62.90

We also computed the regression-based estimate of μ_y, using the formula $\hat{\mu}_{yreg} = \hat{\mu}_{yrss} + \hat{B}(\mu_x - \bar{x}_{rss})$, where (Yu and Lam [173])

$$\hat{B} = [\sum_{r=1}^{5}\sum_{i=1}^{6}(x_{(r)i} - \bar{x}_{rss})(y_{[r]i} - \bar{y}_{rss})]/[\sum_{r=1}^{5}\sum_{i=1}^{6}(x_{(r)i} - \bar{x}_{rss})^2]. \qquad (8.1)$$

Here $\hat{\mu}_{yrss} = \bar{y}_{rss}$. It turns out that $\hat{\mu}_{reg} = 20.00$. Its estimated variance is indeed very small because of the extremely high value of the sample correlation (Theorem 1 in Yu and Lam [173]). These are also reported in Table 8.2.

For the same values of the population parameters, we have generated 1000 more RSS data sets each of size 30 in the usual fashion by first ranking X and then selecting judgment ordered Y, and computed the estimates of μ_y and σ^2_y. Summary statistics of these estimates are given below.

It is clear from the above table that even allowing for sampling fluctuations, RSS does much better than SRS for estimation of the population mean. Of course, use of regression estimates of μ_y would produce even much better results.

8.2 Cost effective gasoline sampling

Traditionally unburned hydrocarbons are emitted from automobile tailpipes to produce ground level ozone and smog. With advances in the automobile technology, many of these hydrocarbons actually evaporate off the manifold. To deal with these evaporative emissions requires controls on both the vehicle and the gasoline. The most effective gasoline control is to reduce the volatility of the gasoline, as commonly measured by its Reid Vapor Pressure (RVP). Regulations limiting the RVP, among others, have led to *reformulated gasoline* appearing in appropriate areas of the US in warmer months. RVP may be measured at the laboratory and also at the gas station itself. While laboratory analysis of RVP is not unduly expensive, it is expensive to ship the field sample of gasoline to the laboratory since the gasoline sample must be packed so as not to allow escaping gaseous hydrocarbons, and special transport measures are used for flammable liquids such as gasoline. Thus, reducing the quantity of lab testing without any major loss of accuracy in RVP measurement would allow large cost savings. Fortunately, cheap field measurements of RVP which are highly correlated covariates for laboratory measurements of RVP enable us to apply the notion of RSS very effectively in this context.

Nussbaum and Sinha [111] studied the above problem. Based on 20 paired values of (field, lab) RVP data denoted by (X, Y), given below in Table 8.4, we can compute the statistics: $\bar{x} = 8.642$, $\bar{y} = 8.649$, $s_x^2 = 0.497$, $s_y^2 = 0.542$, $r_{xy} = 0.97$. Given the extremely high value of the correlation between X and Y, it is clear that X, which is very easy and cheap to measure, can be taken as a covariate to rank Y-values.

We have used the 90 X-values from Table 3 of Nussbaum and Sinha [111] and the fitted linear regression: $Y = -0.12 + 1.015X + \epsilon$ to generate 90 population Y-values, and then we have used both the SRS method and the RSS method to generate a sample of size 15 to compare these methods for estimation of the mean of Y. We have taken the distribution of ϵ to be normal with mean 0 and variance 0.0083, matching with the observed correlation of 0.97 between Y and X from Table 8.4. The population (X, Y)-values are given below in Table 8.5. The 90 Y-values are randomly divided into 10 sets of 3×3 squares, and used to compute the variances of the three order statistics. These are given in Table 8.6 and used later in connection with UBRSS.

In Table 8.7, we have reported the SRS, balanced RSS and unbalanced RSS values. Based on these sample values, we have computed some summary statistics, which are reported in Table 8.8. It is clear from this table that RSS is doing much better than SRS. We should note that the estimated variance 0.00182 of the estimate of the population mean μ_y reported in Table 8.8 in the unbalanced RSS case is obtained by applying the formula:

$$\hat{var}(\hat{\mu}_y(ubrss)) = \frac{1}{k^2} \sum_{r=1}^{k} \frac{\hat{\sigma}_{(r)}^2}{n_r}. \tag{8.2}$$

Table 8.4. Field and Lab RVP data

Pump	Field RVP(X)	Lab RVP(Y)
1	7.60	7.59
2	9.25	9.33
3	9.21	9.30
4	8.89	9.01
5	8.89	9.02
6	8.88	8.92
7	9.14	9.28
8	9.15	9.28
9	8.25	8.60
10	8.98	8.56
11	8.63	8.64
12	8.62	8.70
13	7.90	7.83
14	8.32	7.99
15	8.28	8.03
16	8.25	8.29
17	8.17	8.21
18	10.72	10.67
19	7.85	7.86
20	7.86	7.86

Also, the estimate 0.04662 of the population variance σ_y^2 based on the UBRSS reported in Table 8.8 is obtained from the formula ([132]):

$$\hat{\sigma}^2_{ubrss} = \sum_{r=1}^{k}[(1 + \frac{1}{k(n_r - 1)})\frac{S^2_{(r)}}{kn_r} + \frac{(\bar{y}_{(r)} - \hat{\mu}_y(ubrss))^2}{k}]. \qquad (8.3)$$

We also generated 15 RSS Y-values by first observing X-values from Table 8.5, ranking them and then selecting the corresponding Y-values. To do this, we drew an SRS of size 45 out of 90, divided them randomly into 5 sets of 9 each, then arranged 9 observations in each set into 3×3 square, and then drew RSS sample of X and hence Y-values under both balanced and unbalanced schemes. These values along with some summary statistics are given in Tables 8.9, 8.10. It is again obvious that even in this situation RSS is doing much better than SRS. The amount of sample correlation in this case is 0.9545.

8.3 Estimation of weights of browse and herbage

Halls and Dell [58] reported a very interesting and useful study related to the estimation of mean weights of browse and herbage in a pine-hardwood forest of East Texas, USA to demonstrate the superiority of RSS over SRS. About

Table 8.5. The population (X,Y)-values

No.	X	Y	No.	X	Y	No.	X	Y
1	7.27	7.42043	31	7.47	7.40194	61	7.35	7.30603
2	7.57	7.55349	32	7.54	7.47252	62	7.41	7.54441
3	7.47	7.49534	33	7.54	7.49941	63	7.28	7.17364
4	7.27	7.15787	34	7.45	7.43206	64	7.41	7.37344
5	7.51	7.70336	35	6.42	6.48164	65	7.37	7.37414
6	8.03	7.97076	36	8.21	8.09954	66	7.63	7.54699
7	7.37	7.40452	37	8.69	8.80488	67	7.37	7.40092
8	7.16	7.13687	38	8.64	8.61522	68	7.45	7.29943
9	8.32	8.26775	39	7.86	7.95413	69	7.47	7.53020
10	8.30	8.30437	40	8.22	8.20800	70	7.37	7.43612
11	7.51	7.43280	41	7.35	7.21393	71	7.32	7.33868
12	7.01	6.96980	42	7.37	7.14588	72	7.30	7.31769
13	7.52	7.57230	43	7.41	7.36116	73	7.22	7.10908
14	6.53	6.43238	44	7.45	7.49847	74	7.47	7.41043
15	7.01	6.92487	45	7.44	7.46515	75	7.54	7.58248
16	7.54	7.45800	46	8.34	8.29940	76	7.31	7.36234
17	7.31	7.32792	47	8.56	8.64405	77	7.25	7.36149
18	7.59	7.55373	48	7.32	7.22311	78	7.37	7.26742
19	7.37	7.23511	49	7.35	7.45366	79	7.32	7.23074
20	7.47	7.49863	50	7.50	7.35767	80	7.28	7.33903
21	7.56	7.66931	51	7.47	7.49101	81	7.38	7.49330
22	7.34	7.45694	52	7.37	7.35279	82	7.22	7.11519
23	7.56	7.51281	53	7.43	7.30706	83	7.76	7.69649
24	7.45	7.58576	54	7.41	7.41159	84	7.45	7.42787
25	7.60	7.49504	55	7.37	7.28150	85	7.51	7.56681
26	7.63	7.53542	56	7.31	7.28819	86	7.47	7.54080
27	7.16	7.23144	57	7.59	7.50309	87	7.38	7.33081
28	7.54	7.56595	58	7.47	7.52883	88	7.79	7.77385
29	7.51	7.49295	59	7.43	7.42279	89	7.38	7.46263
30	7.52	7.62099	60	7.40	7.56357	90	7.14	7.03363

Table 8.6. Population variance in each rank

$\sigma^2_{(1)}$	$\sigma^2_{(2)}$	$\sigma^2_{(3)}$
0.01651	0.10463	0.14914

300 acres of land were divided into 126 sets of 3 quadrats each. Under SRS, one quadrat was selected at random from each set and data were collected on both browse and herbage. Under RSS (almost) balanced scheme, 43 sets were used for the highest yield, and in each set the quadrat *expected* to have the highest yield was identified. Likewise, 46 sets were used for the medium yield, and in each set the quadrat *expected* to have the medium yield was identified. Lastly, 37 sets were used for the lowest yield, and in each set the

Table 8.7. SRS, BRSS and UBRSS samples

No.	SRS sample	No.	BRSS sample	No.	UBRSS sample
1	7.32792	1	7.30603	1	7.45800
2	7.58248	2	7.77385	1	7.33868
3	7.95413	3	7.54080	2	7.28819
4	7.37344	1	7.23144	2	7.40452
5	7.33903	2	7.37414	2	7.54699
6	7.54699	3	7.49941	2	7.49330
7	7.55349	1	7.13687	2	7.46263
8	7.51281	2	7.35767	3	7.55349
9	6.96980	3	7.56357	3	7.49534
10	7.49330	1	7.42279	3	8.29940
11	7.49941	2	7.66931	3	7.57230
12	6.48164	3	7.58248	3	7.56357
13	7.52883	1	7.40092	3	7.49295
14	7.49101	2	7.50309	3	7.97076
15	7.45694	3	7.46515	3	7.51281

Table 8.8. Summary statistics

Method	Mean(y)	Variance(y)	Variance of mean(y)
Population	7.48350	0.138029	0.00153
SRS	7.40741	0.10587	0.00706
BRSS	7.45517	$\hat{\sigma}^2_{Stokes}$=0.027195	0.00111
		$\hat{\sigma}_1^2$=0.02718	
		$\hat{\sigma}_2^2$=0.02649	
		$\hat{\sigma}_3^2$=0.02374	
		$\hat{\sigma}_4^2$=0.03543	
UBRSS	7.50668	0.04662	0.00182

quadrat *expected* to have the lowest yield was identified. To apply the RSS unbalanced scheme, Halls and Dell (1966) actually used data from the entire population (total of 378 quadrats) to compute the population mean of 13.1 gms for browse and 7.1gms for herbage, and also the population variances in each category: highest, medium, lowest. This resulted in the standard deviations proportions 4 : 2 : 1 for browse, and 2 : 1 : 1 for herbage. Accordingly, under the unbalanced RSS scheme, 72, 40 and 14 sets were used, respectively, in highest, medium, and lowest category for browse, and 62, 31 and 31 sets were used, respectively, in highest, medium, and lowest category for herbage. The results of statistical analysis appear below. It is obvious that RSS is much superior to SRS.

Table 8.9. SRS, BRSS and UBRSS samples

No.	SRS sample	No.	BRSS sample	No.	UBRSS sample
1	7.77385	1	7.13687	1	7.33903
2	7.47252	2	7.36234	1	7.26742
3	7.53020	3	7.54441	1	7.31769
4	7.40092	1	7.32792	1	7.11519
5	7.69649	2	7.42279	1	7.36149
6	6.48164	3	7.49941	1	6.43238
7	7.55373	1	7.23074	1	7.49330
8	7.23074	2	7.28819	1	7.13687
9	7.03363	3	7.95413	2	7.49847
10	7.49534	1	7.36149	2	7.50309
11	7.49330	2	7.41159	3	8.26775
12	7.33081	3	7.57230	3	7.69649
13	7.52883	1	6.92487	3	7.47252
14	7.42787	2	7.50309	3	8.20800
15	8.61522	3	7.66931	3	7.57230

Table 8.10. Summary statistics

Method	Mean(y)	Variance(y)	Variance of mean(y)
Population	7.48350	0.138029	0.00153
SRS	7.47101	0.19435	0.01296
BRSS	7.41396	$\hat{\sigma}^2_{Stokes}$=0.05661	0.00156
		$\hat{\sigma}^2_1$=0.05353	
		$\hat{\sigma}^2_2$=0.05440	
		$\hat{\sigma}^2_3$=0.05788	
		$\hat{\sigma}^2_4$=0.04310	
UBRSS	7.50904	0.14480	0.00452
REG	7.51902	$\hat{\sigma}^2_{Stokes}$=0.05661	(based on $\hat{\sigma}^2_{Stokes}$) 0.000349
		$\hat{\sigma}^2_1$=0.05353	(based on $\hat{\sigma}^2_1$) 0.000330
		$\hat{\sigma}^2_2$=0.05440	(based on $\hat{\sigma}^2_2$) 0.000336
		$\hat{\sigma}^2_3$=0.05788	(based on $\hat{\sigma}^2_3$) 0.000357
		$\hat{\sigma}^2_4$=0.04310	(based on $\hat{\sigma}^2_4$) 0.000266

8.4 Estimation of shrub phytomass in Appalachian oak forests

Martin et al. [95] undertook an interesting study to again demonstrate the efficiency of RSS over the use of SRS in the context of estimation of shrub phytomass in Appalachian oak forests. The experiment involved four large areas each of $20m \times 20m$ dimension, each such area in turn divided into 16 plots (as 4 rows and 4 columns), and four types of vegetation: mixed hardwood, mixed oak, mixed hardwood and pine, and mixed pine. The four areas were separately devoted to the four vegetation types, and yields from

Table 8.11. Statistics for 126 ranked and unranked browse quadrats

System of sampling quadrats	Mean yields[1]	Variance of mean	Relative number of clipped quadrats for precision approximating that of random sampling
	Grams		
Unranked: random selections (SRS)	14.9	4.55	100
Perfectly ranked, nearly equal numbers in high, medium, and low groups	13.2	2.18	48
Perfectly ranked, numbers in groups nearly proportional to standard deviation (SD)	12.9	1.91	42
Ranked by technical range man, numbers in groups nearly proportional to SD	12.8	1.83[2]	40
Ranked by woodsworker, numbers in groups nearly proportional to SD	13.4	2.04	45

[1]The mean yield of 378 clipped quadrats was 13.1 grams (oven-dry) per quadrat.
[2]Presumably the small inconsistent difference in estimates of the variance for perfect ranking and ranking by technical range man resulted from random variation.

all 64 plots were recorded. Obviously, one can then compute the population mean and population variance for each vegetation type separately, and also the combined mean and variance on the basis of all 64 plots. It turns out that the overall mean is 2536 kg/ha and the variance of the mean is 0.15×10^6 kg/ha.

The statistical analysis of Martin et al. [95] falls into *two* categories: an *actual* implementation of SRS and RSS, and a *simulation* study. Under the first part, an SRS of size 4 plots were selected out of 16 plots from each vegetation area, and observations were recorded and used to compute the overall sample mean of 1976 kg/ha and estimated variance of the overall sample mean as 4.54×10^6 kg/ha . On the other hand, when RSS of set size 4 was used in each vegetation area (balanced), overall mean turned out to be 2356 kg/ha , which is much closer to the true mean, and the estimated variance of the RSS mean was 2.73×10^6 kg/ha, which is much smaller than that of SRS sample mean.

Under the simulation study part, it was decided to work with a sample of size $N = 80$ for each vegetation type. Thus, an SRS of size 1 was repeated 80 times, and the mean and the variance of the mean were computed. Under RSS balanced scheme, the set size was taken as 4 and number of cycles as 20. Under RSS unbalanced scheme, the available population figures were used to determine the appropriate proportions of the four order statistics, and

Table 8.12. Statistics for 124 ranked and unranked herbage quadrats

System of sampling quadrats	Mean yields[1]	Variance of mean	Relative number of clipped quadrats for precision approximating that of random sampling
	Grams		
Unranked: random selections (SRS)	7.3	1.00	100
Perfectly ranked, nearly equal numbers in high, medium, and low groups	7.0	.73	73
Perfectly ranked, numbers in groups nearly proportional to standard deviation (SD)	7.2	.58	58
Ranked by technical range man, numbers in groups nearly proportional to SD	6.9	.53	53
Ranked by woodsworker, numbers in groups nearly proportional to SD	6.9	.53	53

[1]The mean yield of 372 clipped quadrats was 7.1 grams (oven-dry) per quadrat.

Table 8.13. Number of measured plots in each rank* and vegetation type for simulated ranked set samples with unequal allocation

VEGETATION TYPE

Rank	Mixed Hardwood	Mixed Oak	Mixed Oak and Pine	Mixed Pine
1	50	42	62	24
2	18	30	6	20
3	10	6	5	27
4	2	2	7	9

* Highest rank is 1.

these appear in Table 8.13. Once the RSS data were thus generated for each vegetation type, RSS means and estimates of variances of the means were computed in the usual way. These are reported in Table 8.14. It is obvious from this table that RSS is doing much better than SRS.

Table 8.14. Summary of simulated random and ranked set sample results with 80 measured plots within each vegetation type

Sampling Method	Mean Phytomass (kg/ha)	Variance of the Mean (x 10^6)	Coefficient of Variation of the Mean%
		Mixed Hardwood	
All 16 Plots	656	0.03	28
Random	684	0.50	103
RS-Equal Allocation	652	0.06	38
RS-Unequal Allocation	591	0.05	38
		Mixed Oak	
All 16 Plots	1096	0.09	28
Random	1014	1.15	106
RS-Equal Allocation	1021	0.11	33
RS-Unequal Allocation	1025	0.14	37
		Mixed Oak and Pine	
All 16 Plots	1927	0.17	21
Random	1939	2.70	85
RS-Equal Allocation	1727	0.19	25
RS-Unequal Allocation	1826	0.26	28
		Mixed Pine	
All 16 Plots	6467	0.87	14
Random	7012	6.91	37
RS-Equal Allocation	6436	1.78	21
RS-Unequal Allocation	5892	1.38	20

8.5 Single family homes sales price data

In this section we consider 63 single family homes sales price data given in Appendix A and compare the effectiveness of SRS and RSS based on a sample of size 15. The data are obtained from Table 4.2, pages 138-139, of Bowerman and O'Connell [29]. We consider these 63 values as comprising the population so that the population mean is 78.8 and the population variance is 1206.94. Dividing the 63 values at random into 7 sets of 9 each, and then arranging 9 observations within each set into a 3×3 square, we computed $\sigma^2_{(1)} = 199$, $\sigma^2_{(2)} = 763$, and $\sigma^2_{(3)} = 965$, which are in the ratio of $1 : 4 : 5$. An SRS of size 15 is drawn from the population of 63 values, and the sample mean and sample variance are obtained as $\bar{x} = 70.5$ and $s^2 = 1306.5$, resulting in the estimate of the variance of the sample mean as 87. To draw an RSS of size 15, we first draw an SRS of size 45 out of 63 values, then divide the sample data at random into 5 sets of 9 each, then divide the 9 values within each set

Table 8.15. Ranked set sample of single family homes sales price

	i				
r	1	2	3	4	5
1	53.4	44.5	65.0	68.0	34.0
2	60.0	100.0	101.0	52.0	73.0
3	115.0	139.0	69.0	149.0	94.0

Table 8.16. Estimates of population mean and variance of single family homes sales price based on RSS

$\hat{\mu}_{RSS}$	$\hat{var}(\hat{\mu}_{RSS})$	$\hat{\sigma}^2_{Stokes}$	$\hat{\sigma}^2_1$	$\hat{\sigma}^2_2$	$\hat{\sigma}^2_3$	$\hat{\sigma}^2_4$
81.13	39.5	1163.70	1110.32	1125.62	1186.80	926.77

at random into a 3×3 square, and finally we draw an RSS of size 3 from each set. These values $\{X_{(r)i}\}$ are given below in Table 8.15.

Several statistics can now be routinely computed based on the above values. These are given below in Table 8.16. It is clear that RSS is doing much better than SRS for estimation of the population mean.

We also generated 20 sets of RSS from the above population values and computed various summary statistics. These are given below in Table 8.17.

From the above table, it is evident that even allowing for sampling fluctuations, RSS offers much improvement compared to SRS. Some salient features of the sampling distributions of the above statistics are summarized in the following Table 8.18.

8.6 Tree data

In this last section on applications of RSS, we refer to an important data set from Platt et al. [133] related to 399 conifer (pinus palustris) trees. The original data were collected on seven variables of which we have used only two: X, the diameter in centimeters at breast height, and Y, the entire height in feet. To make an easy application of RSS, we deleted the last 3 observations. This truncated data set is reported in Appendix B.

Treating this as the population and considering its random decomposition into 44 squares each of dimension 3×3, we computed the population means $\mu_x = 20.9641$, $\mu_y = 52.6768$, the population variances $\sigma_x^2 = 309.545$, $\sigma_y^2 = 3253.44$, and also the variances of the three order statistics $\sigma^2_{y(1)} = 156.37$, $\sigma^2_{y(2)} = 1140.40$ and $\sigma^2_{y(3)} = 4174.35$ only for Y, which can be used when we consider an unbalanced RSS.

To compare SRS and RSS, we drew an SRS of size $n = 30$ and computed the sample mean and the sample variance for Y. For RSS, we drew a total of 90 simple random sample from the above population, divided these observations

Table 8.17. Estimates of population mean and variance of single family homes sales price for 20 RSS samples

sample	$\hat{\mu}_{RSS}$	$\hat{var}(\hat{\mu}_{RSS})$	$\hat{\sigma}^2_{Stokes}$	$\hat{\sigma}^2_1$	$\hat{\sigma}^2_2$	$\hat{\sigma}^2_3$	$\hat{\sigma}^2_4$
1	81.13	39.50	1163.70	1110.32	1125.62	1186.80	926.77
2	83.16	42.45	1508.82	1436.24	1450.68	1508.43	1263.01
3	85.66	36.32	1417.18	1387.01	1359.02	1247.05	1722.92
4	76.27	30.92	713.75	712.85	697.08	634.03	902.00
5	78.33	19.72	575.90	558.75	557.22	551.12	577.06
6	80.53	55.19	950.89	936.61	942.68	966.97	863.77
7	72.85	28.11	551.74	545.45	543.06	533.52	574.07
8	77.81	53.95	1069.38	1051.76	1052.03	1053.12	1048.51
9	78.69	21.89	1138.41	1079.90	1084.41	1102.42	1025.85
10	76.55	47.60	747.75	746.28	745.50	742.36	755.69
11	88.92	78.02	2346.64	2278.20	2268.21	2228.27	2398.03
12	75.89	25.87	766.12	760.90	740.91	660.97	1000.74
13	85.97	60.33	2194.37	2128.62	2108.40	2027.54	2371.22
14	73.96	29.24	800.31	786.70	776.20	734.21	912.66
15	76	33.62	1059.61	1022.21	1022.59	1024.08	1017.74
16	82.79	74.33	2005.68	1978.58	1946.29	1817.14	2366.02
17	78.68	51.86	1088.85	1063.77	1068.12	1085.52	1011.58
18	72.7	27.12	693.41	665.79	674.30	708.35	563.65
19	86.04	53.62	2353.34	2262.49	2250.07	2200.37	2411.59
20	76.57	96.97	2090.21	2037.16	2047.83	2090.53	1909.08

Table 8.18. Summary statistics of estimates of population mean and variance for 20 RSS samples

variable	Mean	S.D.	Median
$\hat{\mu}_{RSS}$	79.43	4.70	78.51
$\hat{var}(\hat{\mu}_{RSS})$	45.33	20.56	40.97
$\hat{\sigma}^2_{Stokes}$	1261.80	611.86	1079.12
$\hat{\sigma}^2_1$	1227.48	590.82	1057.77
$\hat{\sigma}^2_2$	1223.01	587.10	1060.08
$\hat{\sigma}^2_3$	1205.14	575.26	1069.32
$\hat{\sigma}^2_4$	1281.10	656.40	1014.66

into 10 sets of 9 each, then divided each set at random into a 3×3 square, and finally performed RSS based on X-values, which eventually led to RSS Y-values. For UBRSS, we used the replications $n_1 = 2$, $n_2 = 6$ and $n_3 = 22$ for the three order statistics in the usual way. All these sample values are reported in Table 8.19.

Based on the above sample values, we can easily compute estimates of the population mean and variance of Y as well as the estimates of the variance of the estimated mean of Y by each of the three methods: SRS, BRSS, UBRSS.

Table 8.19. SRS, BRSS and UBRSS samples

No.	SRS sample	No.	BRSS sample	No.	UBRSS sample
1	219	1	8	1	22
2	16	2	38	1	66
3	11	3	103	2	16
4	94	1	5	2	33
5	33	2	40	2	36
6	28	3	223	2	38
7	203	1	6	2	26
8	180	2	70	2	22
9	16	3	78	3	38
10	154	1	4	3	37
11	5	2	27	3	140
12	6	3	232	3	208
13	11	1	21	3	105
14	32	2	16	3	107
15	19	3	96	3	21
16	43	1	26	3	92
17	8	2	82	3	104
18	85	3	106	3	92
19	25	1	11	3	222
20	35	2	12	3	109
21	119	3	91	3	38
22	17	1	20	3	120
23	68	2	40	3	87
24	152	3	176	3	244
25	19	1	6	3	84
26	23	2	9	3	33
27	222	3	38	3	105
28	223	1	6	3	111
29	9	2	11	3	113
30	3	3	154	3	192

These are reported below in Table 8.20. It is obvious from this table that both the BRSS and the UBRSS are doing much better than SRS.

To study the behavior of RSS from a frequentist point of view, we also generated 20 sets of RSS from the above population and computed various summary statistics. These are given below in Table 8.21, 8.22. It is thus clear that even allowing for sampling fluctuations, RSS does much better than SRS.

To use the RSS regression estimate of the mean of Y, we generated afresh 25 SRS on Y and an equal number of BRSS samples on both X and Y, using $k = n = 5$. This is done in the usual way by first selecting RSS X-values and then the corresponding Y-values (judgment ordered), resulting in $(x_{(r)i}, y_{[r]i})$. These values are reported below in Table 8.23. As in Section 8.1 and 8.2, we have used the results from Yu and Lam [173] to compute the regression-based

Table 8.20. Summary statistics

Method	Mean(y)	Variance(y)	Variance of mean(y)
Population	52.68	3253.44	8.22
SRS	69.27	5773.03	192.43
BRSS	58.5	$\hat{\sigma}^2_{Stokes}$=4203.36	53.29
		$\hat{\sigma}^2_1$=4116.50	
		$\hat{\sigma}^2_2$=4116.54	
		$\hat{\sigma}^2_3$=4116.88	
		$\hat{\sigma}^2_4$=4115.50	
UBRSS	60.56	2008.90	74.23

Table 8.21. Estimates of population mean and variance of tree data for 20 RSS samples

sample	μ_{RSS}	var(μ_{RSS})	σ^2_{Stokes}	σ^2_1	σ^2_2	σ^2_3	σ^2_4
1	50.33	78.60	3019.20	2951.76	2997.15	3405.68	1726.18
2	60.17	116.54	5409.11	5311.50	5345.35	5649.92	4397.78
3	55.37	62.55	3250.52	3195.03	3204.71	3291.90	2933.45
4	50.67	71.16	3744.02	3706.69	3690.39	3543.61	4147.02
5	50.97	54.89	2413.27	2403.63	2387.72	2244.61	2832.96
6	52.47	76.41	3279.98	3235.54	3247.06	3350.77	2924.41
7	55.87	83.11	3751.91	3700.50	3709.96	3795.10	3445.08
8	49.53	52.67	2170.88	2149.55	2151.18	2165.87	2105.49
9	48.70	81.93	4016.70	3976.14	3964.74	3862.14	4283.96
10	58.87	106.86	3963.77	3990.92	3938.51	3466.84	5405.93
11	64.80	102.96	4142.30	4117.18	4107.19	4017.21	4387.12
12	60.63	66.88	4742.03	4636.67	4650.84	4778.45	4253.85
13	52.53	49.50	2902.40	2839.61	2855.15	2995.07	2419.87
14	52.53	60.76	2698.46	2679.74	2669.28	2575.18	2962.04
15	49.23	83.24	3321.43	3274.17	3293.95	3471.96	2740.16
16	42.83	41.33	1943.94	1934.88	1920.47	1790.78	2323.94
17	56.67	107.75	3705.20	3691.55	3689.44	3670.45	3748.53
18	42.13	68.84	3375.02	3331.03	3331.36	3334.24	3322.38
19	54.4	81.51	4378.52	4309.94	4314.09	4351.39	4198.04
20	48.17	53.26	2471.87	2430.83	2442.74	2549.86	2109.47

RSS estimate of μ_y and also its estimated variance. Other results are also obtained in a similar way. These are all reported in Table 8.24. As expected, we find the performance of the regression-based RSS estimate of μ_y to be very impressive compared to those of SRS- and BRSS-based estimates.

Table 8.22. Summary statistics of estimates of population mean and variance based on 20 RSS samples

variable	Mean	S.D.	Median
$\hat{\mu}_{RSS}$	52.84	5.71	52.5
$\hat{var}(\hat{\mu}_{RSS})$	75.04	21.11	73.79
$\hat{\sigma}^2_{Stokes}$	3435.03	881.07	3348.23
$\hat{\sigma}^2_1$	3393.34	865.74	3302.60
$\hat{\sigma}^2_2$	3395.56	868.08	3312.66
$\hat{\sigma}^2_3$	3415.55	910.46	3436.26
$\hat{\sigma}^2_4$	3333.38	983.92	3142.21

Table 8.23. SRS and BRSS samples

No.	SRS sample	No.	BRSS sample
1	12	1	(13.2,38)
2	77	2	(3.2,2)
3	41	3	(18.1,21)
4	7	4	(47.9,137)
5	82	5	(45.9,202)
6	9	1	(2.6,4)
7	7	2	(3.1,4)
8	5	3	(18.4,22)
9	176	4	(34,99)
10	34	5	(41.1,105)
11	105	1	(12.5,34)
12	26	2	(13.4,21)
13	3	3	(21.3,40)
14	34	4	(18,82)
15	140	5	(57.8,188)
16	25	1	(2.5,3)
17	26	2	(3.7,6)
18	70	3	(12.7,38)
19	85	4	(36.4,103)
20	10	5	(39.8,196)
21	239	1	(3,4)
22	5	2	(14.1,40)
23	13	3	(19.9,55)
24	34	4	(41.8,92)
25	105	5	(36.7,77)

Table 8.24. Summary statistics

Method	Mean(y)	Variance(y)	Variance of mean(y)
Population	52.68	3253.44	8.22
SRS	54.8	3634.42	145.38
BRSS	64.52	$\hat{\sigma}^2_{Stokes}$=3897.76	36.76
		$\hat{\sigma}^2_1$=3780.18	
		$\hat{\sigma}^2_2$=3778.61	
		$\hat{\sigma}^2_3$=3772.31	
		$\hat{\sigma}^2_4$=3811.69	
REG	59.50	$\hat{\sigma}^2_{Stokes}$=3897.76	(based on $\hat{\sigma}^2_{Stokes}$) 26.69
		$\hat{\sigma}^2_1$=3780.18	(based on $\hat{\sigma}^2_1$) 25.89
		$\hat{\sigma}^2_2$=3778.61	(based on $\hat{\sigma}^2_2$) 25.88
		$\hat{\sigma}^2_3$=3772.31	(based on $\hat{\sigma}^2_3$) 25.83
		$\hat{\sigma}^2_4$=3811.69	(based on $\hat{\sigma}^2_4$) 26.10

Appendix A: Sales price of sixty-three single-family residences

Residence, i	Sales Price, y (×$1000)	Residence, i	Sales Price, y (×$1000)	Residence, i	Sales Price, y (×$1000)
1	53.5	22	87.9	43	90.0
2	49.0	23	80.0	44	83.0
3	50.5	24	94.0	45	115.0
4	49.9	25	74.0	46	50.0
5	52.0	26	69.0	47	55.2
6	55.0	27	63.0	48	61.0
7	80.5	28	67.5	49	147.0
8	86.0	29	35.0	50	210.0
9	69.0	30	142.5	51	60.0
10	149.0	31	92.2	52	100.0
11	46.0	32	56.0	53	44.5
12	38.0	33	63.0	54	55.0
13	49.5	34	60.0	55	53.4
14	105.0	35	34.0	56	65.0
15	152.5	36	52.0	57	73.0
16	85.0	37	75.0	58	40.0
17	60.0	38	93.0	59	141.0
18	58.5	39	60.0	60	68.0
19	101.0	40	73.0	61	139.0
20	79.4	41	71.0	62	140.0
21	125.0	42	83.0	63	55.0

Appendix B: Tree data

No.	Diameter	Height	No.	Diameter	Height	No.	Diameter	Height
1	15.9	28.0	34	4.7	14.0	67	3.8	8.0
2	22.0	26.0	35	11.0	19.0	68	41.2	94.0
3	56.9	119.0	36	58.8	222.0	69	39.8	68.0
4	9.6	16.0	37	3.5	4.0	70	18.6	33.0
5	24.6	43.0	38	10.1	28.0	71	38.7	68.0
6	3.3	7.0	39	16.9	38.0	72	12.2	17.0
7	11.4	21.0	40	10.8	26.0	73	6.0	16.0
8	4.7	6.0	41	9.0	21.0	74	8.0	14.0
9	21.3	40.0	42	8.0	19.0	75	13.5	19.0
10	16.8	28.0	43	17.8	38.0	76	20.1	32.0
11	5.1	12.0	44	23.9	37.0	77	57.4	202.0
12	7.5	22.0	45	2.3	5.0	78	8.2	22.0
13	3.1	7.0	46	5.8	13.0	79	32.7	41.0
14	4.9	7.0	47	6.0	16.0	80	9.4	23.0
15	6.1	9.0	48	8.8	23.0	81	8.9	25.0
16	5.5	12.0	49	9.9	20.0	82	9.2	18.0
17	6.5	11.0	50	14.6	34.0	83	6.1	14.0
18	5.6	14.0	51	10.8	29.0	84	7.5	19.0
19	6.9	11.0	52	44.2	181.0	85	52.3	152.0
20	3.8	6.0	53	12.9	16.0	86	15.5	25.0
21	9.7	27.0	54	28.0	77.0	87	23.7	51.0
22	6.9	16.0	55	39.8	76.0	88	67.1	208.0
23	4.1	8.0	56	20.4	37.0	89	12.3	16.0
24	58.5	192.0	57	47.3	111.0	90	14.0	16.0
25	46.0	203.0	58	35.7	66.0	91	4.9	9.0
26	22.2	51.0	59	44.9	87.0	92	5.5	8.0
27	3.7	5.0	60	8.7	25.0	93	7.6	17.0
28	52.9	162.0	61	24.3	46.0	94	3.5	5.0
29	63.2	223.0	62	15.7	35.0	95	6.3	18.0
30	46.5	211.0	63	30.9	54.0	96	19.0	39.0
31	56.3	196.0	64	69.2	131.0	97	2.7	5.0
32	21.9	43.0	65	24.1	72.0	98	8.2	24.0
33	11.0	20.0	66	4.2	8.0	99	7.6	20.0

Tree data (continued)

No.	Diameter	Height	No.	Diameter	Height	No.	Diameter	Height
100	9.2	27.0	133	12.4	31.0	166	4.7	8.0
101	5.9	9.0	134	15.1	34.0	167	5.3	10.0
102	6.2	12.0	135	12.7	38.0	168	10.6	19.0
103	13.3	22.0	136	49.0	96.0	169	3.7	6.0
104	13.4	30.0	137	20.8	35.0	170	3.9	8.0
105	33.9	82.0	138	11.9	18.0	171	5.3	12.0
106	33.7	93.0	139	47.6	154.0	172	2.5	3.0
107	8.3	26.0	140	10.6	32.0	173	13.2	38.0
108	48.0	99.0	141	22.9	33.0	174	17.1	37.0
109	40.4	78.0	142	10.6	27.0	175	13.9	33.0
110	8.6	22.0	143	49.7	103.0	176	8.0	21.0
111	16.0	26.0	144	50.6	122.0	177	8.5	27.0
112	29.1	49.0	145	19.1	40.0	178	50.1	109.0
113	18.4	22.0	146	53.0	114.0	179	6.8	18.0
114	26.8	37.0	147	18.0	82.0	180	19.9	55.0
115	6.2	7.0	148	44.4	105.0	181	17.5	47.0
116	2.9	6.0	149	10.8	35.0	182	6.8	21.0
117	3.0	8.0	150	51.7	219.0	183	10.9	33.0
118	14.6	20.0	151	22.6	48.0	184	11.2	23.0
119	18.4	32.0	152	7.7	19.0	185	20.2	38.0
120	15.0	34.0	153	43.5	60.0	186	19.6	26.0
121	18.4	41.0	154	3.1	3.0	187	18.4	46.0
122	44.5	64.0	155	5.0	13.0	188	50.9	84.0
123	4.5	8.0	156	4.4	8.0	189	17.6	42.0
124	10.4	20.0	157	3.3	5.0	190	44.1	113.0
125	24.0	37.0	158	2.6	5.0	191	17.0	31.0
126	5.1	10.0	159	53.5	211.0	192	46.9	135.0
127	5.3	13.0	160	48.9	206.0	193	2.8	6.0
128	2.5	4.0	161	47.8	176.0	194	25.5	40.0
129	2.2	3.0	162	17.2	37.0	195	14.5	28.0
130	3.1	4.0	163	28.6	45.0	196	14.1	40.0
131	2.6	4.0	164	10.8	31.0	197	47.1	85.0
132	8.1	26.0	165	50.1	212.0	198	42.2	93.0

Tree data (continued)

No.	Diameter	Height	No.	Diameter	Height	No.	Diameter	Height
199	40.2	75.0	232	17.2	24.0	265	17.5	46.0
200	66.8	223.0	233	57.0	213.0	266	8.9	33.0
201	4.1	11.0	234	6.3	9.0	267	47.4	53.0
202	60.6	180.0	235	44.2	216.0	268	22.0	49.0
203	8.0	15.0	236	3.0	4.0	269	6.8	18.0
204	17.2	43.0	237	36.4	62.0	270	7.5	18.0
205	22.0	46.0	238	2.7	3.0	271	22.2	32.0
206	15.9	39.0	239	4.4	7.0	272	19.3	25.0
207	3.1	4.0	240	41.4	177.0	273	14.5	22.0
208	4.5	12.0	241	3.4	7.0	274	3.5	5.0
209	32.0	65.0	242	8.4	25.0	275	10.9	26.0
210	46.9	126.0	243	4.8	12.0	276	14.7	33.0
211	36.4	103.0	244	4.2	5.0	277	12.5	34.0
212	25.4	64.0	245	6.3	16.0	278	18.7	35.0
213	40.0	82.0	246	32.6	67.0	279	20.5	38.0
214	40.4	87.0	247	15.3	31.0	280	11.5	26.0
215	19.8	42.0	248	38.6	42.0	281	43.7	92.0
216	30.5	37.0	249	5.2	6.0	282	10.1	36.0
217	37.7	183.0	250	61.8	239.0	283	42.1	70.0
218	22.1	33.0	251	10.9	33.0	284	41.8	92.0
219	5.5	6.0	252	3.5	6.0	285	21.9	70.0
220	28.4	76.0	253	2.5	4.0	286	56.9	113.0
221	46.4	120.0	254	10.9	26.0	287	40.5	83.0
222	15.8	33.0	255	8.9	24.0	288	15.9	76.0
223	45.9	202.0	256	21.0	67.0	289	18.8	58.0
224	33.5	82.0	257	44.1	107.0	290	26.5	89.0
225	36.7	77.0	258	7.0	16.0	291	42.2	133.0
226	44.0	105.0	259	9.4	27.0	292	39.8	196.0
227	51.6	197.0	260	8.0	17.0	293	48.2	197.0
228	45.0	78.0	261	23.0	59.0	294	25.5	40.0
229	34.0	99.0	262	11.6	35.0	295	19.6	40.0
230	53.1	198.0	263	33.0	90.0	296	59.4	176.0
231	30.8	85.0	264	7.5	17.0	297	9.3	25.0

Tree data (continued)

No.	Diameter	Height	No.	Diameter	Height	No.	Diameter	Height
298	19.8	33.0	331	38.8	91.0	364	27.0	29.0
299	34.0	42.0	332	41.1	105.0	365	19.9	24.0
300	4.9	6.0	333	39.0	116.0	366	17.5	22.0
301	8.3	14.0	334	45.4	140.0	367	62.5	232.0
302	3.7	8.0	335	47.9	137.0	368	44.6	92.0
303	32.7	53.0	336	53.7	105.0	369	38.0	167.0
304	2.6	7.0	337	43.5	96.0	370	3.2	2.0
305	44.8	140.0	338	18.7	68.0	371	13.4	21.0
306	10.3	21.0	339	57.8	188.0	372	5.7	14.0
307	28.5	32.0	340	14.9	23.0	373	3.6	5.0
308	34.0	119.0	341	4.5	10.0	374	2.6	3.0
309	36.6	81.0	342	8.8	22.0	375	75.4	244.0
310	50.8	106.0	343	23.6	26.0	376	2.2	5.0
311	29.2	78.0	344	11.5	21.0	377	3.7	7.0
312	8.5	21.0	345	20.0	27.0	378	3.1	4.0
313	23.4	35.0	346	8.3	19.0	379	7.2	26.0
314	7.9	15.0	347	12.6	30.0	380	8.2	20.0
315	44.6	149.0	348	5.8	14.0	381	3.2	5.0
316	2.5	4.0	349	12.9	25.0	382	2.5	3.0
317	9.4	17.0	350	5.4	11.0	383	4.0	11.0
318	3.0	6.0	351	22.5	42.0	384	1.8	1.0
319	2.8	3.0	352	11.8	32.0	385	2.7	3.0
320	3.0	5.0	353	51.2	203.0	386	9.9	21.0
321	4.1	8.0	354	45.3	85.0	387	6.3	11.0
322	23.4	42.0	355	48.7	120.0	388	3.2	11.0
323	59.0	189.0	356	6.6	20.0	389	3.3	5.0
324	5.2	8.0	357	16.7	33.0	390	5.0	12.0
325	8.5	10.0	358	12.3	23.0	391	3.7	6.0
326	7.8	15.0	359	6.5	15.0	392	2.0	5.0
327	44.9	140.0	360	53.0	106.0	393	5.1	13.0
328	54.4	104.0	361	18.1	21.0	394	6.0	12.0
329	47.9	129.0	362	2.4	5.0	395	3.8	8.0
330	41.3	94.0	363	5.8	11.0	396	3.5	9.0

References

1. W. Abu-Dayyeh and H. A. Muttlak. Using ranked set sampling for hypothesis tests on the scale parameter of the exponential and uniform distributions. *Pakistan Journal of Statistics*, 12:131–138, 1996.
2. M. F. Al-Saleh and A. I. Al-Omari. Multistage ranked set sampling. *Journal of statistical planning and inference*, 102:273–286, 2002.
3. M. F. Al-Saleh, K. Al-Shrafat, and H. A. Muttlak. Bayesian estimation using ranked set sampling. *Biometrical Journal*, 42:489–500, 2000.
4. M. F. Al-Saleh and G. Zheng. Estimation of bivariate characteristics using ranked set sampling. *Australian & New Zealand Journal of Statistics*, 44:221–232, 2002.
5. E. Aragon, G. P. Patil, and C. Taillie. A performance indicator for ranked set sampling using ranking error probability matrix. *Environmental and Ecological Statistics*, 6:75–80, 1999.
6. Z.D. Bai and Z. Chen. On the theory of ranked set sampling and its ramifications. *Journal of Statistical Planning and Inference*, 109:81–99, 2003.
7. L. Barabesi. The computation of the distribution of the sign test statistics for ranked set sampling. *Communications in Statistics. Simulation and Computation*, 27:833–842, 1998.
8. L. Barabesi. The unbalanced ranked-set sample sign test. *J. Nonparametr. Statist.*, 13:279–289, 2001.
9. L. Barabesi and A. Al-Sharaawi. The efficiency of ranked set sampling for parameter estimation. *Statistics & Probability Letters*, 53:189–199, 2001.
10. L. Barabesi and L. Fattorini. Kernel estimators for the intensity of a spatial point process by point-to-plant distances: Random sampling vs. ranked set sampling. Technical Report TR 94-0402, Center for Statistical Ecology and Environmental Statistics. Department of Statistics Penn State University. University Park. PA., 1994.
11. L. Barabesi and L. Fattorini. Kernel estimators of probability density functions by ranked-set sampling. *Commun. Statist.— Theory Meth.*, 31(4):597–610, 2002.
12. L. Barabesi and M. Marcheselli. Ranked set sampling in sample surveys with auxiliary information. *Environmental and Ecological Statistics*.
13. L. Barabesi and C. Pisani. Steady-state ranked set sampling for replicated environmental sampling designs. *Environmentrics*.

14. L. Barabesi and C. Pisani. Ranked set sampling for replecated sampling designs. *Biometrics*, 58:586–592, 2002.
15. V. Barnett. Ranked set sample design for environmental investigations. *Environmental and Ecological Statistics*, 6:59–74, 1999.
16. V. Barnett and K. Moore. Best linear unbiased estimates in ranked set sampling with particular reference to imperfect ordering . *Journal of Applied Statistics*, 24:697–710, 1997.
17. M. C. M. Barreto and V. Barnett. Best linear unbiased estimators for the simple linear regression model using ranked-set sampling. *Environmental and Ecological Statistics*, 6:119–134, 1999.
18. P. K. Bhattacharya. Induced order statistics: theory and applications. In P. R. Krishnaiah and P. K. Sen, editors, *Handbook of Statistics*, volume 4, pages 383–403. Elsevier Science Publishers, 1984.
19. D. S. Bhoj. Estimation of parameters of the extreme value distribution using ranked set sampling. *Communications in Statistics - Theory and Methods*, 26:653–, 1997.
20. D. S. Bhoj and M. Ahsanullah. Estimation of parameters of the generalized geometric distribution using ranked set sampling. *Biometrics*, 52:685–694, 1996.
21. D.S. Bhoj. Estimation of parameters of generalized geometric distribution using ranked set sampling. *Biometrics*, 52:685–694, 1996.
22. D.S. Bhoj. New parametric ranked set sampling. *Journal of Applied Statistical Science*, 6:275–289, 1997.
23. S. J. M. Blaber, D. A. Milton, S. R. Chenery G., and Fry. *New insights into the life-history of Tenualosa ilisha and fishery implications*, 2001.
24. L. L. Bohn. A two-sample procedure for ranked-set samples. Technical Report TR 426, Department of Statistics University of Florida Gainesville. FL., 1993.
25. L. L. Bohn. A review of nonparametric ranked set sampling methodology. *Communications in Statistics - Theory and Methods*, 25:2675–, 1996.
26. L. L. Bohn. A ranked-set sample signed-rank statistic. *J. Nonparametric Statist.*, 9:295–306, 1998.
27. L. L. Bohn and D. A. Wolfe. Non-parametric two sample procedures for ranked set sampling data. *J. Amer. Statist. Assoc.*, 87:552–561, 1992.
28. L. L. Bohn and D. A. Wolfe. The effect of imperfect judgment rankings on properties of procedures based on the ranked-set samples analog of the Mann-Whitney-Wilcoxon statistic. *J. Amer. Statist. Assoc*, 89:168–176, 1994.
29. B. L. Bowerman and R. T. O'Connell. *Forecasting and Time Series.* Duxbury Press, Belmont, California, 1993.
30. R. A. Boyles and F. J. Samaniego. Estimating a distribution function based on nomination sampling. *Journal of the American Statistical Association*, 81:1039–1045, 1986.
31. H. Chen and Z. Chen. *Asymptotic properties of the remedian.* Department of Mathematics, Bowling Green State University and Department of Statistics & Applied Probability, National University of Singapore, 2002.
32. S. H. Chen. Ranked set sampling theory with selective probability vector. *Journal of Statistical Planning and Inference*, 8:161–174, 1983.
33. Z. Chen. Analysis of component reliability of k-out-of-n systems with censoring . *Journal of Statistical Planning and Inference.*
34. Z. Chen. Density estimation using ranked-set sampling data. *Environmental and Ecological Statistics*, 6:135–146, 1999.

35. Z. Chen. On ranked-set sample quantiles and their applications. *Journal of Statistical Planning and Inference*, 83:125–135, 2000.
36. Z. Chen. The efficiency of ranked-set sampling relative to simple random sampling under multi-parameter families. *Statistica Sinica*, 10:247–263, 2000.
37. Z. Chen. Non-parametric inferences based on general unbalanced ranked-set samples. *J. Nonparametr. Statist.*, 13:291–310, 2001.
38. Z. Chen. Ranked-set sampling with regression type estimators . *Journal of Statistical Planning and Inference*, 92:181–192, 2001.
39. Z. Chen. The optimal ranked-set sampling scheme for inference on population quantiles. *Statist. Sinica*, 11:23–37, 2001.
40. Z. Chen. Adaptive ranked set sampling with multiple concomitant variables: an effective way to observational economy. *Bernoulli*, 8:313–322, 2002.
41. Z. Chen and Z. D. Bai. The optimal ranked-set sampling scheme for parametric families. *Sankhyā Ser. A*, 62:178–192, 2000.
42. Z. Chen and L. Shen. Two-layer ranked set sampling with concomitant variables. *Journal of Statistical Planning and Inference*, 115:45–57, 2003.
43. Z. Chen and B. K. Sinha. *Tests for population mean based on a ranked set sample.* Department of Statistics and Applied Probability, National University of Singapore, 2000.
44. Z. Chen and Y. Wang. *Optimal sampling strategies using ranked sets with applications in fish aging and lung cancer studies.* Department of Statistics & Applied Probability, National University of Singapore, 2002.
45. T-J. Cheng, D. C. Christiani, J. K. Wiencke, J. C. Wain, X. Xu, and K. T. Kelsey. Comparison of sister chromatid exchange frequency in peripheral lymphocytes in lung cancer cases and controls. *Mutation Research*, 348:75–82, 1995.
46. T-J. Cheng, D. C. Christiani, X. Xu, , J. C. Wain J. K. Wiencke, and K. T. Kelsey. Increased micronucleus frequency in lymphocytes from smokers with lung cancer. *Mutation Research*, 349:43–50, 1996.
47. N. N. Chuiv and B. K. Sinha. On some aspects of ranked set sampling in parametric estimation. In Rao and Balakrishnan, editors, *Handbook of Statistics.* 1998.
48. J. M. Cobby, M. S. Ridout, and P. J. Bassett R. V. Large. An investigation into the use of ranked set sampling on grass and grass-clover swards. *Grass and Forage Science*, 40:257–263, 1985.
49. H. A. David and D.N. Levine. Ranked set sampling in the presence of judgment error. *Biometrics*, 28:553–555, 1972.
50. T.R. Dell and J. L. Clutter. Ranked set sampling theory with order statistics background. *Biometrika*, 28:545–555, 1972.
51. E. J. Duell, J. K. Wiencke, T-J. Cheng, A. Varkonyi, Z. Zuo, T. D. S. Ashok, E. J. Mark, J. C. Wain, D. C. Christiani, and K. T. Kelsey. Polymorphisms in the DNA repair genes XRCC1 and ERCC2 and biomarkers of DNA damage in human blood mononuclear cells. *Carciongenesis*, 21 (7):1457–1457, 2000.
52. B. Efron. Bootstrap methods: another look at the jackknife. *Ann. Statist.*, 7:1–26, 1979.
53. E. El-Neweihi and B. K. Sinha. Reliability estimation based on ranked set sampling. *Comm. Statist. Theory Methods*, 29:1583–1595, 2000.
54. M. J. Evans. *Application of ranked-set sampling to regeneration surveys in areas direct-seeded to longleaf pine.* PhD thesis, Louisiana State University and Agricultural and Mechanical College, 1967.

55. H. Fei, B. K. Sinha, and Z. Wu. Estimation of parameters in two-parameter Weibull and extreme-value distributions using ranked set sampling. *Journal of Statistical Research*, 28:149–161, 1994.
56. R. O. Gilbert and L. L. Eberhardt. An evaluation of double sampling for estimating plutonium inventory in surface soil. In C. E. Cushing, editor, *Radioecology and Energy Sources*, pages 157–163. Stroudsburg, Pennsylvania: Dowden, Hutchison and Ross, 1976.
57. S. D. Gore, G. P. Patil, A. K. Sinha, and C. Taillie. Certain multivariate considerations in ranked set sampling and composite sampling designs. In G. P. Patil and C. R. Rao, editors, *Multivariate Environmental Statistics*, pages 121–148. North Holland. Amsterdam, 1993.
58. L.S. Halls and T. R. Dell. Trial of ranked set sampling for forage yields. *Forest Science*, 12:22–26, 1966.
59. B A. Hartlaub and D.A. Wolfe. Distribution-free ranked-set sample procedures for umbrella alternatives in the m-sample setting. *Environmental and Ecological Statistics*, 6:105–118, 1999.
60. T.P. Hettmansperger. The ranked set sample sign test. *Journal of Nonparametric Statistics*, 4:263–270, 1995.
61. S.S. Hossain and A. Muttlak. Paired ranked set sampling: a more efficient procedure. *Environmetrics*, 10:195–212, 1999.
62. S.S. Hossain and Hassan A. Muttlak. Selected ranked set sampling. *Australian and New Zealand Journal of Statistics.*
63. J. Huang. Asymptotic properties of the NPMLE of a ditribution based on ranked set samples. *The Annals of Statistics*, 25:1036–1049, 1997.
64. W. Huang, L. M. Ryan, S. W. Thurston, K. T. Kelsey, J. K. Wiencke, and D. C. Christiani. *A Likelihood-based Model for Using Multiple Markers in Environmental Health*, 2001.
65. G. D. Johnson and G. P. Patil. Ranked set sampling: A novel method for increasing efficiency of natural resource assessment and environmental monitoring. Technical Report FR 94-0203, Center for Statistical Ecology and Environmental Statistics. Department of Statistics. Penn State University University Park. PA, 1994.
66. G. D Johnson, G.P. Patil, and A.K. Sinha. Ranked set sampling for vegetation research. *Abstracta Botanica*, 17:87–102, 1993.
67. A. Kaur, J. Orsin, and G. P. Patil. Ranked set sample sign test for quantiles under unequal allocation. Technical Report TR 96-0603, Center for Statistical Ecology and Environmental Statistics, Department of Stattstics. Penn State University University Park, PA, 1990.
68. A. Kaur, G. P. Patil, S. J. Shirk, and C. Taillie. Ranked set sampling: an annotated bibliography. *Environmental and Ecological Statistics*, 2:25–54, 1995.
69. A. Kaur, G. P. Patil, S. J. Shirk, and C. Taillie. Environmental sampling with a concomitant variable: a comparison between ranked set sampling and stratified simple random sampling. *Journal of Applied Statistics*, 23:231–255, 1996.
70. A. Kaur, G. P. Patil, and C. Taillie. Optimal allocation for symmetric distributions in ranked set sampling. *Annals of the Institute of Statistical Mathematics.*
71. A. Kaur, G. P. Patil, and C. Taillie. Ranked set sample sign test under unequal allocation. *Journal of Statistical Planning and Inference.*
72. A. Kaur, G. P. Patil, and C. Taillie. Optimal allocation models for symmetric distributions in ranked set sampling. Technical Report TR 95-0201, Center

for Statistical Ecology and Environmental Statistics. Department of Statistics. Penn State University. University Park PA, 1995.

73. A. Kaur, G. P. Patil, and C. Taillie. Precision versus cost: A comparison between ranked set sampling and stratified simple random sampling. Technical Report TR 95-0101, Center for Statistical Ecology and Environmental Statistics Department of Statistics. Penn State University. University Park. PA, 1995.
74. A. Kaur, G. P. Patil, and C. Taillie. Unequal allocation models for ranked set sampling with skew distributions. *Biometrics*, 53:123–130, 1997.
75. A. Kaur, G. P. Patil, C. Taillie, and J. Wit. Ranked set sample sign test for quantiles. *Journal of Statistical Planning and Inference.*
76. D. H. Kim and Y. C. Kim. Wilcoxon signed rank test using ranked-set sample. *Korean J. Comput. Appl. Math.*, 3:235–243, 1996.
77. Y. H. Kim. *Estimation of a distribution function under generalized ranked set sampling.* PhD thesis, University of California, Graduate Program in Applied Statistics. University of California. Riverside. CA, 1995.
78. Y. H. Kim and B. C. Arnold. Parameter estimation under generalized ranked set sampling. *Statistics and Probability Letters.*
79. K. M. Koti and G. J. Babu. Sign test for ranked set sampling. *Communications in Statistics - Theory and Methods*, 25:1617–1630, 1996.
80. E. Kouider and H. Charif. Mode ranked set sampling. *Far East J. Theor. Stat.*, 5:143–157, 2001.
81. P. R. Krishnaiah and P. K. Sen. Tables for order statistics. In P. R. Krishnaiah and P. K. Sen, editors, *Handbook of Statistics*, volume 4, pages 892–897. Elsevier Science Publishers, 1984.
82. P. H. Kvam and F. J. Samaniego. On maximum likelihood estimation based on ranked set sampling with applications to reliability. In A. Basu, editor, *Advances in Reliability*, pages 215–229. 1993.
83. P. H. Kvam and F. J. Samaniego. On the inadmissibility of empirical averages as estimators in ranked set sampling. *Journal of Statistical Planning and Inference*, 36:39–55, 1993.
84. P. H. Kvam and F. J. Samaniego. *On the inadmissibility of standard estimators based on ranked set sampling*, pages 291 -292. 1993.
85. P. H. Kvam and F. J. Samaniego. Nonparametric maximum likelihood estimation based on ranked set samples. *Journal of the American Statistical Associations*, 89:526–537, 1994.
86. P. H. Kvam and R. C. Tiwari. Bayes estimation of a distribution function using ranked set samples. *Environmental and Ecological Statistics*, 6:11–12, 1999.
87. H. Lacayo, N. K. Neerchal, and B. K. Sinha. Ranked set sampling from a dichotomous population. *Journal of Applied Statistical Science.*
88. K. F. Lam, P. L. H. Yu, and C. F. Lee. Kernel method for the estimation of the distribution function and the mean with auxiliary information in ranked set sampling. *Environmetrics*, 13:397–406, 2002.
89. M. Lavine. The Bayesics of ranked set sampling. *Environmental and Ecological Statistics*, 6:47–57, 1999.
90. D. Li. *On some applications of ranked set sampling in statistical inference.* PhD thesis, University of Maryland, 1997.
91. D. Li and N. Chuiv. On the efficiency of ranked set sampling strategies in parametric estimation. *Calcutta Statistical Association Bulletin*, 47:23–42, 1997.
92. D. Li, B. K. Sinha, and N. Chuiv. On estimation of $P(X > c)$ based on a ranked set sample. *Essays in Statistics.*

93. D. Li, B. K. Sinha, and F. Perron. Random selection in ranked set sampling and its applications. *Journal of Statistical Planning and Inference.*
94. S. N. MacEachern, Ömer Öztürk, D. A. Wolfe, and G. V. Stark. A new ranked set sample estimator of variance. *J. R. Statist. Soc. B*, 64:177–188, 2002.
95. W. L. Martin, T. L. Shank, R. G. Oderwald, and D. W. Smith. Evaluation of ranked set sampling for estimating shrub phytomass in Appalachian Oak forest. Technical Report No. FWS-4-80, School of Forestry and Wildlife Resources VPI&SU Blacksburg, VA, 1980.
96. G. A. McIntyre. A method for unbiased selective sampling using ranked sets. *Australian Journal of Agricultural Research*, 3:385–390, 1952.
97. G. A. McIntyre. Statistical aspects of vegetation sampling. In L . T. Mannetje, editor, *Measurement of Grassland Vegetation and Animal Production*, pages 8–21. 1978.
98. H. A. Muttlak. *Some aspects of ranked set sampling with size biased probability selection.* PhD thesis, University of Wyoming Laramic, 1988.
99. H. A. Muttlak. Combining the line intercept sampling with the ranked set sampling. *Journal of Information and Optimization Sciences*, 16:1–, 1995.
100. H. A. Muttlak. Parameters estimation in a simple linear regression using rank set sampling. *The Biometrical Journal*, 37:799–810, 1995.
101. H. A. Muttlak. Estimation of parameters for one-way layout with rank set sampling . *The Biometrical Journal*, 38:407–515, 1996.
102. H. A. Muttlak. Pair ranked set sampling. *The Biometrical Journal*, 38:879–885, 1996.
103. H. A. Muttlak. Median ranked set sampling . *Journal of Applied Statistical Science*, 6, 1997.
104. H. A. Muttlak. Median ranked set sampling with concomitant variables and a comparison with ranked set sampling and regression estimators. *Environmetrics*, 9:255–267, 1998.
105. H. A. Muttlak and L. L. McDonald. Ranked set sampling with respect to concomitant variables and with size biased probability of selection. *Communications in Statistics - Theory and Methods*, 19:205–219, 1990.
106. H. A. Muttlak and L. L. McDonald. Ranked set sampling with size biased probability of selection. *Biometrics*, 46:435–445, 1990.
107. H. A. Muttlak and L. L. McDonald. Ranked set sampling and the line intercept method: A more efficient procedure. *Biometrical Journal*, 34:329–346, 1992.
108. R. W. Nahhas, D. A. Wolfe, and H. Chen. Ranked set sampling: cost and optimal set size. *Biometrics*, 58:964–971, 2002.
109. R. C. Norris. Estimation of multiple characteristics by ranked set sampling methods. Master's thesis, Department of Statistics, Penn State University, University Park PA, 1994.
110. R.C. Norris, G. P. Patil, and A. K. Sinha. Estimation of multiple characteristics by ranked set sampling methods. *Coenoses*, 10:95–111, 1995.
111. B.D. Nussbaum and B. K. Sinha. Cost effective gasoline sampling using ranked set sampling. In *American Statistical Association 1997 Proceedings of the Section on Statistics and the Environment*, pages 83–87. American Statistical Association, August 1997.
112. Ömer Öztürk. One- and two-sample sign tests for ranked set samples with selective designs. *Commun. in Statist. Theory and methods*, 28:1231–1245, 1999.

113. Ömer Öztürk. Two-sample inference based on one-sample ranked set sample sign statistics. *Journal of Nonparametric Statistics*, 10:197–212, 1999.
114. Ömer Öztürk. Rank regression in ranked-set samples. *Journal of the American Statistical Association*, 97:1180–1191, 2002.
115. Ömer Öztürk. Ranked-set sample inference under a symmetry restriction. *Journal of Statistical Planning and Inference*, 102:317–336, 2002.
116. Ömer Öztürk and D. A. Wolfe. Optimal ranked set sampling protocol for the signed rank test. Technical Report TR 630, Ohio State University Department of Statistics, 1998.
117. Ömer Öztürk and D. A. Wolfe. Alternative ranked set sampling protocols for the sign test. *Statist. Probab. Lett.*, 47:15–23, 2000.
118. Ömer Öztürk and D. A. Wolfe. An improved ranked set two-sample Mann-Whitney-Wilcoxon test. *Can. J. Statist.*, 28:123–135, 2000.
119. Ömer Öztürk and D. A. Wolfe. Optimal allocation procedure in ranked set sampling for unimodal and multi-modal distributions. *Environmental and Ecological Statistics*, 7:343–356, 2000.
120. Ömer Öztürk and D. A. Wolfe. A new ranked set sampling protocol for the signed rank test. *J. Statist. Plann. Inference*, 96:351–370, 2001.
121. Ross N. P. and Stokes L. Editorial: Special issue on statistical design and analysis with ranked set samples. *Environmental and Ecological Statistics*, 6:5–9, 1999.
122. G. P. Patil. Editorial: Ranked set sampling. *Environmental and Ecological Statistics*, 2:271–285, 1995.
123. G. P. Patil, A. K. Sinha, and C. Taillie. Ranked set sampling and ecological data analysis. Technical Report TR 91-1203, Center for Statistical Ecology and Environmental Statistics, Department of Statistics Penn State University, University Park, PA, 1991.
124. G. P. Patil, A. K. Sinha, and C. Taillie. Ranked set sampling from a finite population in the presence of a trend on a site. *Journal of Applied Statistical Sciences*, 1:51–65, 1993.
125. G. P. Patil, A. K. Sinha, and C. Taillie. Relative precision of ranked set sampling: Comparison with the regression estimator. *Environmetrics*, 4:399–412, 1993.
126. G. P. Patil, A. K. Sinha, and C. Taillie. Ranked set sampling. In G. P. Patil and C. R. Rao, editors, *Handbook of Statistics Volumn 12: Environmental Statistics*, pages 167–200. 1994.
127. G. P. Patil, A. K. Sinha, and C. Taillie. Ranked set sampling for multiple characteristics. *International Journal of Ecology and Environmental Sciences*, 20:357–373, 1994.
128. G. P. Patil, A. K. Sinha, and C. Taillie. Finite population corrections for ranked set sampling. *Annals of the Institute of Statistical Mathematics*, 47:621–636, 1995.
129. G. P. Patil, A. K. Sinha, and C. Taillie. Ranked set sampling coherent rankings and size-biased permutations. *Journal of Statistical Planning and Inference*, 63:311 -324, 1997.
130. G. P. Patil, A. K. Sinha, and C. Taillie. Ranked set sampling: a bibliography. *Environmental and Ecological Statistics*, 6:91–98, 1999.
131. G. P. Patil and C. Taillie. Environmental sampling, observational economy and statistical inference with emphasis on ranked set sampling encounter sampling

and composite sampling. In *Bulletin of the International Statistical Institute*, pages 295–312. 1993.

132. F. Perron and B. K. Sinha. Estimation of variance based on a ranked set sample. *Journal of Statistical Planning and Inference*.
133. W.J. Platt, G. M. Evans, and S.L. Rathbun. The population dynamics of a long-lived conifer(Pinus palustris). *American Naturalist*, 131:491–525, 1988.
134. B. Presnell and L. L. Bohn. U-statistics and imperfect ranking in ranked set sampling. *Journal of Nonparametric Statistics*, 10:111–126, 1999.
135. M. S. Ridout and J. M. Coby. Ranked set sampling with non-random selection of sets and errors in ranking. *Applied Statistics*, 36:145–152, 1987.
136. N. Risch and H. Zhang. Extreme discordant sib pairs for mapping quantitative trait loci in humans. *Science*, 268:1584–1589, 1995.
137. S. Rodriguez. *Unequal allocation in ranked set sampling*. PhD thesis, Penn State University, 1994.
138. P. J. Rousseeuw and G. W. Jr. Bassett. The remedian: a robust averaging method for large data sets. *Journal of the American Statistical Association*, 85:97–104, 1990.
139. H. M. Samawi. Stratified ranked set sample. *Pakistan Journal of Statistics*, 12:9–16, 1996.
140. H. M. Samawi. *On quantiles estimation with application to normal ranges and Hodges-Lehmann estimate using a variety of ranked set sample*. Department of Statistics, Yarmouk University, Irbid, Jordan, 1999.
141. H. M. Samawi, M. S. Ahmed, and W. Ahu-Dayyeh. Estimating the population mean using extreme ranked set sampling. *Biometrical Journal*, 38:577–587, 1996.
142. H. M. Samawi and Omar A. M. Al-Sagheer. On the estimation of the distribution function using extreme and median ranked set sampling. *Biometrocal Journal*, 43:357–373, 2001.
143. H. M. Samawi and H. A. Muttlak. Estimation of ratio using rank set sampling. *Biometrical Journal*, 38:753–764, 1996.
144. S. Sarikavanij, M. Tiensuwan, and B. K. Sinha. Some novel applications of ranked set sampling. Technical report, Department of Mathematics and Statistics, University of Maryland and Department of Mathematics, Mahidol University, Thailand, 2003.
145. J.M. Sengupta, I. M. Chakravarti, and D. Sarkar. Experimental survey for the estimation of cinchona yield. *Bulletin of the International Statistical Institute*, 33 (2):313–331, 1951.
146. R. J. Serfling. *Approximation theorems of mathematical statistics*. John Wiley & Sons, New York, 1980.
147. W. H. Shen. Use of ranked set sampling for test of a normal mean. *Calcutta Statistical Association Bulletin*, 44:183–193, 1994.
148. W.H. Shen and W. Yuan. A test for a normal mean based on a modified partial ranked set sample. *Pakistan Journal of Statistics*, 11:227–233, 1995.
149. S. Shirahata. An extension of the ranked set sampling theory . *Journal of Statistical Planning and Inference*, 6:65–72, 1982.
150. S. Shirahata. Interval estimation in ranked set sampling. *Bulletin of the Computational Statistics of Japan*, 6:15–, 1993.
151. S. J. Shirk. *Sampling with a concomitant variable: A comparison between ranked set sampling and stratified random sampling*. PhD thesis, Department

of Statistics, Center for Statistical Ecology and Environmental Statistics, Penn State University, 1994.
152. P. L. D. N. Silva and C. J. Skinner. Variable selection for regression estimation in finite populations. *Survey Methodology*, 23:23–32, 1997.
153. B. W. Silverman. *Density estimation for statistics and data analysis.* Chapman and Hall, London, New York, 1986.
154. A. K. Sinha, J. E. Rodriguez, and G. P. Patil. A comparison of the McIntyre's ranked set sample estimator with the regression estimator. Technical Report TR 94-0303, Center for Statistical Ecology and Environmental Statistics Department of Statistics Penn State University. University Park PA, 1995.
155. B. K. Sinha, B. K. Sinha, and S. Purkayastha. On some aspects of ranked set sampling for estimation of normal and exponential parameters. *Statistics and Decisions*, 14:223–240, 1996.
156. S. L. Stokes. *An investigation of the consequences of ranked set sampling.* PhD thesis, University of North Carolina, Chapel Hill NC, 1976.
157. S. L. Stokes. Ranked set sampling with concomitant variables. *Communications in Statistics - Theory and Methods*, A6(12):1207–1211, 1977.
158. S. L. Stokes. Inferences on the correlation coefficient in bivariate normal populations from ranked set samples. *Journal of the American Statistical Association*, 75:989–995, 1980.
159. S. L. Stokes. Estimation of variance using judgment ordered ranked set samples. *Biometrics*, 36:35–42, 1980.
160. S. L. Stokes. Ranked set sampling. In S. Kott, N. L. Johnson, and C. B. Read, editors, *Encyclopedia of Statistical Sciences Vol. 7*, pages 385–388. 1986.
161. S. L. Stokes. Parametric ranked set sampling. *Annals of the Institute of Statistical Mathematics*, 47:465–482, 1995.
162. S. L. Stokes and T. W. Sager. Characterization of a ranked set sample with application to estimating distribution functions. *Journal of the American Statistical Association*, 83:374–381, 1988.
163. K. Takahasi. On the estimation of the population mean based on ordered samples from an equicorrelated multivariate distribution. *Annals of the Institute of Statistical Mathematics*, 21:249–255, 1969.
164. K. Takahasi. Practical note on estimation of poulation means based on samples stratified by means of ordering. *Annals of the Institute of Statistical Mathematics*, 22:421–428, 1970.
165. K. Takahasi and M. Futatsuya. Ranked set samplimig from a finite population (Japanese). *Proceedings of the Institute of Statistical Mathematics*, 36:55–68, 1988.
166. K. Takahasi and M. Futatsuya. Dependence between order statistics in samples from finite population and its application to ranked set sampling. *Annals of the Institute of Statistical Mathematics*, 50:49–70, 1998.
167. K. Takahasi and K. Wakimoto. On unbiased estimates of the population mean based on the sample stratified by means of ordering. *Annals of the Institute of Statistical Mathematics*, 20:1–31, 1968.
168. R. C. Tiwari. Estimation of a distribution function under nomination sampling. *IEEE Transaction on Reliability*, 35:558–561, 1988.
169. R. C. Tiwari and M. T. Wells. Quantile estimation based on nomination sampling. *IEEE Transaction on Reliability*, 38:612–614, 1989.
170. T. R. Willemain. Estimating the population median by nomination sampling. *Journal of the American Statistical Association*, 75:908–911, 1980.

171. T. Yanagawa and S. Chen. The MG-procedure in ranked set sampling. *Journal of Statistical Planning and Inference*, 4:33–44, 1980.
172. T. Yanagawa and S. Shirahata. Ranked set sampling theory with selective probability matrix. *Australian Journal of Statistics*, 19:45–52, 1976.
173. P. L. H. Yu and K. Lam. Regression estimator in ranked set sampling. *Biometrics*, 53:1070–1080, 1997.
174. P. L. H. Yu, K. Lam, and B. K. Sinha. Estimation of variance based on balanced and unbalanced ranked set samples. *Environmental and Ecological Statistics*, 6:23–46, 1999.
175. X. Zhao and Z. Chen. On the Ranked-Set Sampling M-Estimates for Symmetric Location Families. *Annals of the Institute of Statistical Mathematics*, 54:626–640, 2002.

Index

Lecture Notes in Statistics

For information about Volumes 1 to 121, please contact Springer-Verlag

122: Timothy G. Gregoire, David R. Brillinger, Peter J. Diggle, Estelle Russek-Cohen, William G. Warren, and Russell D. Wolfinger (Editors), Modeling Longitudinal and Spatially Correlated Data. x, 402 pp., 1997.

123: D. Y. Lin and T. R. Fleming (Editors), Proceedings of the First Seattle Symposium in Biostatistics: Survival Analysis. xiii, 308 pp., 1997.

124: Christine H. Müller, Robust Planning and Analysis of Experiments. x, 234 pp., 1997.

125: Valerii V. Fedorov and Peter Hackl, Model-Oriented Design of Experiments. viii, 117 pp., 1997.

126: Geert Verbeke and Geert Molenberghs, Linear Mixed Models in Practice: A SAS-Oriented Approach. xiii, 306 pp., 1997.

127: Harald Niederreiter, Peter Hellekalek, Gerhard Larcher, and Peter Zinterhof (Editors), Monte Carlo and Quasi-Monte Carlo Methods 1996. xii, 448 pp., 1997.

128: L. Accardi and C.C. Heyde (Editors), Probability Towards 2000. x, 356 pp., 1998.

129: Wolfgang Härdle, Gerard Kerkyacharian, Dominique Picard, and Alexander Tsybakov, Wavelets, Approximation, and Statistical Applications. xvi, 265 pp., 1998.

130: Bo-Cheng Wei, Exponential Family Nonlinear Models. ix, 240 pp., 1998.

131: Joel L. Horowitz, Semiparametric Methods in Econometrics. ix, 204 pp., 1998.

132: Douglas Nychka, Walter W. Piegorsch, and Lawrence H. Cox (Editors), Case Studies in Environmental Statistics. viii, 200 pp., 1998.

133: Dipak Dey, Peter Müller, and Debajyoti Sinha (Editors), Practical Nonparametric and Semiparametric Bayesian Statistics. xv, 408 pp., 1998.

134: Yu. A. Kutoyants, Statistical Inference For Spatial Poisson Processes. vii, 284 pp., 1998.

135: Christian P. Robert, Discretization and MCMC Convergence Assessment. x, 192 pp., 1998.

136: Gregory C. Reinsel, Raja P. Velu, Multivariate Reduced-Rank Regression. xiii, 272 pp., 1998.

137: V. Seshadri, The Inverse Gaussian Distribution: Statistical Theory and Applications. xii, 360 pp., 1998.

138: Peter Hellekalek and Gerhard Larcher (Editors), Random and Quasi-Random Point Sets. xi, 352 pp., 1998.

139: Roger B. Nelsen, An Introduction to Copulas. xi, 232 pp., 1999.

140: Constantine Gatsonis, Robert E. Kass, Bradley Carlin, Alicia Carriquiry, Andrew Gelman, Isabella Verdinelli, and Mike 142: György Terdik, Bilinear Stochastic Models and Related Problems of Nonlinear Time Series Analysis: A Frequency Domain West (Editors), Case Studies in Bayesian Statistics, Volume IV. xvi, 456 pp., 1999.

141: Peter Müller and Brani Vidakovic (Editors), Bayesian Inference in Wavelet Based Models. xiii, 394 pp., 1999.

Approach. xi, 258 pp., 1999.

143: Russell Barton, Graphical Methods for the Design of Experiments. x, 208 pp., 1999.

144: L. Mark Berliner, Douglas Nychka, and Timothy Hoar (Editors), Case Studies in Statistics and the Atmospheric Sciences. x, 208 pp., 2000.

145: James H. Matis and Thomas R. Kiffe, Stochastic Population Models. viii, 220 pp., 2000.

146: Wim Schoutens, Stochastic Processes and Orthogonal Polynomials. xiv, 163 pp., 2000.

147: Jürgen Franke, Wolfgang Härdle, and Gerhard Stahl, Measuring Risk in Complex Stochastic Systems. xvi, 272 pp., 2000.

148: S.E. Ahmed and Nancy Reid, Empirical Bayes and Likelihood Inference. x, 200 pp., 2000.

149: D. Bosq, Linear Processes in Function Spaces: Theory and Applications. xv, 296 pp., 2000.

150: Tadeusz Caliński and Sanpei Kageyama, Block Designs: A Randomization Approach, Volume I: Analysis. ix, 313 pp., 2000.

151: Håkan Andersson and Tom Britton, Stochastic Epidemic Models and Their Statistical Analysis. ix, 152 pp., 2000.

152: David Ríos Insua and Fabrizio Ruggeri, Robust Bayesian Analysis. xiii, 435 pp., 2000.

153: Parimal Mukhopadhyay, Topics in Survey Sampling. x, 303 pp., 2000.

154: Regina Kaiser and Agustín Maravall, Measuring Business Cycles in Economic Time Series. vi, 190 pp., 2000.

155: Leon Willenborg and Ton de Waal, Elements of Statistical Disclosure Control. xvii, 289 pp., 2000.

156: Gordon Willmot and X. Sheldon Lin, Lundberg Approximations for Compound Distributions with Insurance Applications. xi, 272 pp., 2000.

157: Anne Boomsma, Marijtje A.J. van Duijn, and Tom A.B. Snijders (Editors), Essays on Item Response Theory. xv, 448 pp., 2000.

158: Dominique Ladiray and Benoît Quenneville, Seasonal Adjustment with the X-11 Method. xxii, 220 pp., 2001.

159: Marc Moore (Editor), Spatial Statistics: Methodological Aspects and Some Applications. xvi, 282 pp., 2001.

160: Tomasz Rychlik, Projecting Statistical Functionals. viii, 184 pp., 2001.

161: Maarten Jansen, Noise Reduction by Wavelet Thresholding. xxii, 224 pp., 2001.

162: Constantine Gatsonis, Bradley Carlin, Alicia Carriquiry, Andrew Gelman, Robert E. Kass Isabella Verdinelli, and Mike West (Editors), Case Studies in Bayesian Statistics, Volume V. xiv, 448 pp., 2001.

163: Erkki P. Liski, Nripes K. Mandal, Kirti R. Shah, and Bikas K. Sinha, Topics in Optimal Design. xii, 164 pp., 2002.

164: Peter Goos, The Optimal Design of Blocked and Split-Plot Experiments. xiv, 244 pp., 2002.

165: Karl Mosler, Multivariate Dispersion, Central Regions and Depth: The Lift Zonoid Approach. xii, 280 pp., 2002.

166: Hira L. Koul, Weighted Empirical Processes in Dynamic Nonlinear Models, Second Edition. xiii, 425 pp., 2002.

167: Constantine Gatsonis, Alicia Carriquiry, Andrew Gelman, David Higdon, Robert E. Kass, Donna Pauler, and Isabella Verdinelli (Editors), Case Studies in Bayesian Statistics, Volume VI. xiv, 376 pp., 2002.

168: Susanne Rässler, Statistical Matching: A Frequentist Theory, Practical Applications, and Alternative Bayesian Approaches. xviii, 238 pp., 2002.

169: Yu. I. Ingster and Irina A. Suslina, Nonparametric Goodness-of-Fit Testing Under Gaussian Models. xiv, 453 pp., 2003.

170: Tadeusz Caliński and Sanpei Kageyama, Block Designs: A Randomization Approach, Volume II: Design. xii, 351 pp., 2003.

171: David D. Denison, Mark H. Hansen, Christopher C. Holmes, Bani Mallick, and Bin Yu (Editors) Nonlinear Estimation and Classification. viii, 474 pp., 2003.

172: Sneh Gulati and William J. Padgett, Parametric and Nonparametric Inference from Record-Breaking Data. viii, 111 pp., 2003.

173: Jesper Møller (Editor), Spatial Statistics and Computational Methods. xiv, 202 pp., 2003.

174: Yasuko Chikuse, Statistics on Special Manifolds. xxvi, 399 pp., 2003.

175: Jürgen Gross, Linear Regression. Xiv, 394 pp., 2003.

176: Zehua Chen, Zhidong Bai, and Bimal K. Sinha, Ranked Set Sampling: Theory and Application. xii, 224 pp., 2004.

ALSO AVAILABLE FROM SPRINGER!

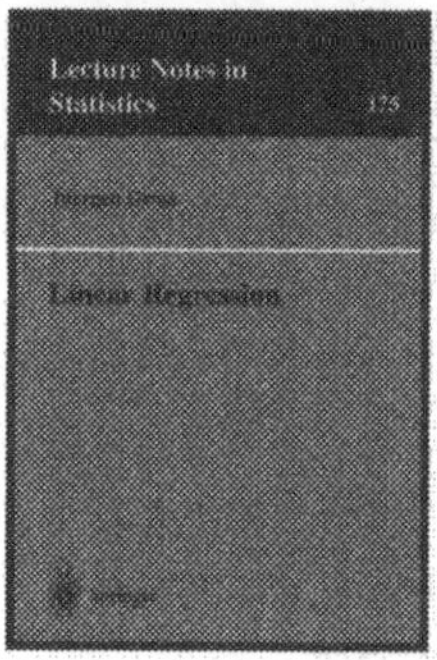

LINEAR REGRESSION

JÜRGEN GROSS

The book covers the basic theory of linear regression models and presents a comprehensive survey of different estimation techniques as alternatives and complements to least squares estimation. The relationship between different estimators is clearly described and categories of estimators are worked out in detail. Proofs are given for the most relevant results, and the presented methods are illustrated with the help of numerical examples and graphics. Special emphasis is laid on the practicability, and possible applications are discussed. The book is rounded off by an introduction to the basics of decision theory and an appendix on matrix algebra.

2003/394 PAGES/SOFTCOVER/ ISBN 3-540-40178-4
LECTURE NOTES IN STATISTICS, VOL. 175

STATISTICS ON SPECIAL MANIFOLDS

YASUKO CHIKUSE

This book is concerned with statistical analysis on the two special manifolds, the Stiefel manifold and the Grassmann manifold, treated as statistical sample spaces consisting of matrices. Statistical analysis on these manifolds is required, especially for low dimensions in practical applications, in the earth (or geological) sciences, astronomy, medicine, biology, meteorology, animal behavior and many other fields. The Grassmann manifold is a rather new subject treated as a statistical sample space, and the development of statistical analysis on the manifold contributes to the related sciences.

2002/418 PAGES/SOFTCOVER/ ISBN 0-387-00160-3
LECTURE NOTES IN STATISTICS, VOL. 174

SPATIAL STATISTICS AND COMPUTATIONAL METHODS

JESPER MØLLER

Spatial statistics and Markov Chain Monte Carlo (MCMC) techniques have each undergone major developments in the last decade. Also, these two areas are mutually reinforcing, because MCMC methods are often necessary for the practical implementation of spatial statistical inference, while new spatial stochastic models in turn motivate the development of improved MCMC algorithms. This volume shows how sophisticated spatial statistical and computational methods apply to a range of problems of increasing importance for applications in science and technology.

2002/214 PAGES/SOFTCOVER/ ISBN 0-387-00136-0
LECTURE NOTES IN STATISTICS, VOL. 173

To Order or for Information:

In the Americas: **CALL:** 1-800-SPRINGER or **FAX:** (201) 348-4505 • **WRITE:** Springer-Verlag New York, Inc., Dept. S5629, PO Box 2485, Secaucus, NJ 07096-2485 • **VISIT:** Your local technical bookstore • **E-MAIL:** orders@springer-ny.com

Outside the Americas: **CALL:** +49 (0) 6221 345-217/8 • **FAX:** + 49 (0) 6221 345-229 • **WRITE:** Springer Customer Service, Haberstrasse 7, 69126 Heidelberg, Germany • **E-MAIL:** orders@springer.de

PROMOTION: S5629

Springer
www.springer-ny.com

9 780387 40263